A PHILOSOPHER LOOKS AT SCIENCE

by

JOHN G. KEMENY

President, Dartmouth College

D. VAN NOSTRAND COMPANY

New York Cincinnati Toronto London Melbourne

Van Nostrand Company Regional Offices:
New York *Cincinnati* *Millbrae*

Van Nostrand Company International Offices:
London *Toronto* *Melbourne*

Library of Congress Catalog Card Number: 59-8064
ISBN: 0-442-04324-4

Published by D. Van Nostrand Company
450 West 33rd Street, New York, N. Y. 10001

Published simultaneously in Canada by
Van Nostrand Reinhold Ltd.

15 14 13 12 11

To Albert Einstein

The author wishes to thank the numerous colleagues, students, and relatives whose severe criticism over a period of years has sharpened the author's views. He also wishes to thank a long list of typists who contributed to the book by their patient understanding. He particularly wishes to thank Mr. S. G. Korenman, who not only criticized an early draft, but also provided the suggestions on which the lists of suggested readings are based.

Table of Contents

Introduction

ONCE UPON A TIME there was an ugly little caterpillar. All the other small animals strutted around, preening their colorful feathers or showing off their glittering coats, while the little caterpillar hid and felt ashamed. Then one day he made up his mind he would not rest until he changed himself into the most beautiful caterpillar in the world. He struggled, he puffed, he almost burst himself trying, but he did succeed. "Look at me," he shouted, "I am truly a lovely caterpillar." But the other animals snickered and laughed at him behind his back. Finally a wise old owl, who had been watching from above, said to the deflated little caterpillar: "The others are not laughing at you because you are not beautiful. Don't you know there is no such thing as a beautiful caterpillar? You have turned yourself into a butterfly."

Every philosopher must learn the lesson of this fable. No matter how hard the philosopher tries to discover the laws of nature, no philosopher can ever do so for the simple reason that, if he succeeds, people will call him a scientist.

It is often pointed out that all of Science grew out of Philosophy. If you read about ancient Greece, you will find a group of philosophers asking questions, the answers to which form the basis of our Science. For a long time they could do no more than ask questions and indulge in more or less ingenious guesses, but slowly the modern scientific method developed, giving definite and well-founded answers to these questions. Slowly the philosopher found himself in a dilemma. If he asked a question, he was a philosopher; if he answered it, he was a scientist.

This book is not a science book. There are many excellent and easily readable books explaining the results of modern Science.

This is a book on the Philosophy of Science. By this time the question will arise: "Are there any questions left for the philosopher to answer?" There certainly are. Roughly speaking, the philosopher deals with those questions that the scientist either does not answer or cannot answer. Fortunately, some of the most interesting questions fall into these categories. But I have restricted myself further; not only is this a philosophy book, but it is a philosophy of science book. Therefore I must restrict myself to such philosophical questions as arise in connection with Science.

In other words, this is not a science book, but a book about Science. This is not nearly so subtle a distinction as one might suppose. A beautiful painting is a piece of art; yet a book explaining the technique used by the artist is *about* Art, not necessarily a piece of art itself. Similarly, if we perform an experiment and write up the results, we are acting as scientists. But when we discuss the general problems involved in experimentation, we are writing about Science. What are these problems that scientists do not write about (unless, of course, they happen to be acting as philosophers)? They include discussions on the scientific method, on the concepts and basic assumptions of Science, its subject matter and its limitations, what it makes use of and what it is used for in turn. In general, it is an attempt to give a unified picture of the nature of Science.

Let me try to illustrate this distinction in terms of the Theory of Relativity. If we ask what this theory can tell us about the motion of planets, that is a scientific question; if we ask why this answer is accepted by scientists, this is a question about scientific method and hence belongs to the Philosophy of Science. If we want to see how the theory is deduced from a few basic assumptions, we go to a science book (and there is certainly none better than Einstein's own account). But if we want to know how these basic assumptions could possibly be justified, we had better ask a philosopher. If we want to learn about mathematical methods used in Physics, any good science department will provide us with the answer, but it is not their job to explain just what the relation is between Mathematics and Science. We would also do well not to ask the average scientist such questions, because it may very well be that he is not in a position to answer us. His whole life is devoted to a task that is all-absorbing and may be more than one human being can accomplish. This is perhaps the reason why many of these tremendously important and interesting questions are left for the philosopher to answer. It is questions of this type that we will try to discuss, and whenever possible we will try to answer them in this book.

Since my basic purpose is to present a unified picture of Science, it is difficult to divide the book into parts. However, I can tell you the basic pattern that I have followed. I start out with certain questions that are presupposed by Science. From this I proceed to a discussion of Science itself and I end up with certain problems that arise out of Science. The first chapter deals with the problem of language and its relevance to the various questions discussed in the book. From this we pass to a discussion of Mathematics, the language that has been found most useful for Science. Since it has often been maintained that the usage of this language involves some basic assumptions, these supposed assumptions are discussed in the third chapter. We are then almost ready to go on to a discussion of Science itself. Just one more tool has to be treated, namely, probabilities. The discussion of Science proper starts with a chapter on the scientific method. After that, four basic questions are asked, questions which arise in a discussion of the scientific method, and a chapter is devoted to each one. This leads us to Chapter 10, which is a discussion of what Science is. Then there are the questions that arise out of Science. Do we live in a completely determined universe? What is Life? What is the nature of our minds? What is the status of values? What is the nature of the social sciences? With this we are brought to the final chapter, which attempts to summarize what has gone before.

As the plan of the book might indicate, it is organized to appeal to the interested layman. I do hope, however, that the unified picture here presented, and the approach of the third chapter (on which the remainder of the book is based), will be of interest even to the expert. A unified picture in so small a space must necessarily be sketchy, so I included a few suggestions for additional reading, which will be found at the end of each chapter. I have tried to arrange these so that references can be quickly found and so that it is easy to see which reference will provide the answer to the question in the reader's mind. If used as an introductory text, it should be supplemented with a good popular book on Science. If it is used in a more advanced course, it should be used in conjunction with some of the books mentioned at the ends of the various chapters. This book could provide the continuity, and the references would fill in details on which the book is too sketchy.

Just one more word of warning. We have already noted the fact that philosophers ask many more questions than they can answer. I believe that asking a good clear question is one of the most important things we can do. We find many instances in the history of both Science and Philosophy where a question was unanswered for centuries until some genius came along and rephrased the question, and all of a sudden it was found that the answer was very simple to find as well. For this reason a great deal of time is spent in this book in clarifying issues. Very often this is the best that I can do.

The book is dedicated to the belief that clarification of a difficult problem is a great step forward. It certainly avoids much fruitless and apparently endless debate, and hence clears the air for fruitful work and the solution of the problem. Nevertheless we are often forced to face one unanswerable question. Have we really learned much? So much that we want to know is still left open, and we remain faced with so many uncertainties about what we do know. Perhaps what we have learned so far is really very little. But when we consider that the pursuit of knowledge for its own sake, the attempt to answer these fundamental questions, has been one of the greatest driving forces for all intellectual pursuits, then it is comforting to note that in all probability these questions, or at least many of them, will forever remain unanswered.

PART ONE

What Science Presupposes

1

Language

"That's a great deal to make one word mean," Alice said in a thoughtful tone.
"When I make a word do a lot of work like that," said Humpty Dumpty, "I always pay it extra."

IT HAS BEEN pointed out that if we had no language by means of which to convey our thoughts and store our knowledge, we would be little different from the lower animals. I would like to add to this, however, that if we had no language we would have no misunderstandings. Every author who writes on a very complex topic is faced with a dilemma. If he writes simply, he is likely to be misunderstood. If he takes the greatest care not to be misunderstood by making his material formidable, he may not be read at all. Since we are going to discuss a highly complex subject matter, and since this is in a field which has been plagued with many linguistic difficulties, I will try to start out by taking a look at some of the hurdles we will have to jump.

PROBLEMS OF LANGUAGE

The real danger in language lies in the fact that we fall into verbal traps a hundred times each day. When did you last say that a book was suggestive? When did you last argue about democracy? Are you in favor of freedom? Do you describe yourself as an Idealist or a Realist? If you are a normal person, you use such words without inhibition many times a day. Yet these words are dangerous. They are dangerous because their meaning is not clear.

The commonest type of verbal sin is using words that are too vague. Vagueness arises because it is much too difficult to define all our words precisely. We use words with a shadowy fringe when it is not clear whether the words apply or do not apply. Of course in many conversations this makes little difference, but the danger comes when we hide our ignorance under the shield of vague words. Suppose a friend asks us how we liked a certain book, and we answer by saying, "It is not really good, but it is very suggestive." How much does our friend know about our real opinion? How bad is 'not really good'? Just exactly what is it that the book suggests? Are we really conveying an idea to him, or have we just used words that are so vague that at some future time we can interpret these in any way whatsoever? Or again, suppose we assert that we are in favor of freedom. This carries some meaning with it, but certainly it means quite different things to different people. Freedom, to the Russian peasant, is a good deal different from the meaning implied by a member of the British House of Lords. In philosophical arguments vague words are extremely popular. This way we convey a general feeling for what we mean, but we do not become sufficiently precise so that the error in our reasoning could be detected. This, of course, is a dangerous practice.

The second kind of danger lies in words that mean too much. The word "democracy" may have an absolutely precise meaning in a given context and may have a second, entirely different, absolutely clear meaning in the next paragraph. Unfortunately this word has been used to mean too many different things. In other words, it is ambiguous. (While I am on the subject, I would like to point out that the word "ambiguous" is itself ambiguous. It can mean either a word having two different meanings or a word having two or more different meanings. This sometimes leads to a good deal of confusion.) Consider classifications of people as Idealists or Realists. Two people talking to each other may be very pleased to find out that they are both Idealists, and yet the first one may mean that he is a World Federalist and the second person may mean that he is an Idealist because he believes that chairs and tables are nothing but the figment of someone's imagination. If we look up the word "realist" in one of the abridged editions of Webster's dictionary, we will find seven different meanings, but it is a safe bet that in actual usage several dozen different meanings have been given to this word.

The third and perhaps most difficult hurdle to overcome is that words arouse emotions. There is a great deal of difference in referring to an Asiatic nation as backward or calling it underdeveloped. Similarly there are various races or religions which, on the one hand, have nice names, and, on the other hand, have names that serve a double purpose as curse words. It would be an interesting exercise to try to take a speech, say, praising our fine public school system, substitute synonyms for synonyms—without changing the meanings of words, only their emotional overtone—and change the original speech into a violent attack on our backward school system. A philosopher is supposed to be free of emotions, and hence philosophical discussions are supposed to avoid such words. In practice, of course, this is impossible.

If ordinary language is full of pitfalls, even for the simple conversations we carry on with our friends, it is certainly dangerous for discussions in Science or philosophical discussions about Science. For this reason, something must be done to construct a new type of language for either or both of these fields. Hence we must discuss the two problems of a language for Science and a language for the Philosophy of Science. Since the solution will turn out to be quite different in the two cases, we must face the problems one at a time.

THE DEFENSE OF VAGUE LANGUAGE

Before we consider various attempts at overcoming the vagueness of ordinary language, we must ask ourselves the question: Is vagueness something desirable or undesirable? We are agreed that ordinary language is full of vagueness, but there are many people who maintain that this is by no means a shortcoming of languages. We must not act like the surgeon who was so anxious to demonstrate his new method of performing an appendectomy that he took out the appendix of a patient suffering from pneumonia.

Some of the arguments in favor of vagueness tell us that, if we limit ourselves to a list of words that are precisely defined and listed in a given dictionary, this would put unnecessary bounds on our imagination. With a limited number of words we can express only a limited number of ideas. This argument, however, rests on a misunderstanding of the nature of language. It is certainly *not* true that with a limited number of words one can express only a limited number of ideas. For example, there is a system known as the binary system of numbers, in which two words, zero and one, suffice to express any number whatsoever, and hence an infinite number of ideas. While I will certainly admit that this system is inconvenient for many purposes, a more familiar example will present a convenient system for doing this. In our decimal notation, just ten different digits suffice to express any number whatsoever quite conveniently. If we allow ourselves a few thousand precisely defined words, we should be able to express almost any idea clearly.

It might be objected that this example is taken from Mathematics, and not from ordinary language. However, there are just as many good examples taken from ordinary language which, incidentally, will indicate that the borderline between it and Mathematics is not so easy to draw. Let us take colors and heat as examples. A medieval painter would certainly have insisted that we will never get a name for every color in the world but will always have to resort to such extremely vague words as "greenish-blue." He would have insisted that you could always paint some shade which would have no exact name. Yet modern science has overcome this difficulty. In terms of wave length, we can give an absolutely precise description of every single shade conceivable. Similarly for heat. If we try to think back to the status of civilization a few thousand years ago, we must certainly imagine that descriptions of how hot something could be were extremely vague. We might have had words like "hot" or "cold." We might have some in-between words for "lukewarm," but nothing approaching a precise description was possible. Yet, with the invention of temperature scales, we have acquired a precise way by which two people can communicate with each other about various stages of hotness and coldness.

I will freely admit that the examples chosen so far were selected to make the case as strong as possible. With most words that are vague and possibly ambiguous we see no way whatsoever of making them clear. But this is a difficulty in practice, not in principle. We have every hope that we will succeed in finding a way for making any given word precise.

There is a final argument in favor of vague language that is difficult to deny. We are told that the type of vagueness found in ordinary languages is necessary for poetry and literature in general. After all, isn't Philosophy itself literature? Of course here we are confronted with a terrible verbal quibble. A more fruitful way of asking the question is this: Isn't it possible for a philosopher's book to be written in such a way that it is a piece of art? This has been done in the past and it is certainly worthwhile whenever it can be done. But the primary purpose of Philosophy is to present ideas and clear arguments. Therefore its fundamental aim must be to clarify thoughts, and its attempt to beautify them must be secondary. The question is also raised whether the poets themselves have not been excellent philosophers. Let us allow Plato to answer this particular question (through the voice of Socrates): "I am almost ashamed to speak of this, but still I must say that there is hardly a person present who would not have talked better about their poetry than they (the poets) did themselves. That showed me in an instant that not by wisdom do poets write poetry, but by a sort of genius and inspiration. They are like diviners or soothsayers, who also say many fine things, but do not understand the meaning of them." This passage is from the *Apology*. The danger hinted at in this passage—of being carried away with the beauty of words—is one that the philosopher must not fall into. Let us grant to the poet and the literary man the right to use vague language, but to the philosopher we cannot give similar permission. He, as well as the scientist, must try to make his thoughts as precise as possible and try to bring beauty into his writings only in so far as this is consistent with his fundamental aim.

THE LANGUAGE OF SCIENCE

Science has solved the problem of finding a suitable language. It has found a language which is perfectly precise and rich enough for all its needs. This language is Mathematics. It is true that many people find this language a difficult one to master, but as long as a sufficient minority of people can speak this language well, it will continue to serve admirably for all the needs of modern Science.

Why is it that Mathematics is so useful as a language for Science? First of all, there is its universality. The most moving sonnet by Keats is wasted on someone who does not speak English, our religious symbols are meaningless to one not schooled in our religious tradition, and even a Beethoven symphony may not appeal to the average Chinese. But every halfway civilized person will realize that two plus two equals four. Mathematics is the only truly universal achievement of man so far.

Bertrand Russell has pointed out that the reason school children have so much trouble understanding what x means is that it doesn't mean anything. This is also the reason why x can be given so many different meanings. This abstract nature of mathematics gives it two more great advantages. First of all, it gives it very wide applicability within Science, and, secondly, it frees Mathematics from one of the greatest evils of ordinary languages, namely, emotional overtones.

These few brief remarks serve only to indicate that Science has solved its problem of language. The way in which Mathematics fulfills this role will be discussed in detail in the next chapter. For the time being, we will have to face the linguistic question confronting the philosopher.

THE LANGUAGE OF PHILOSOPHY

There are many contemporary philosophers who hope that a precise language can be found for Philosophy by means of modern Mathematical Logic. Some remarkable work has been done in this direction, notably by Rudolph Carnap, but at the moment this seems like a utopian undertaking.

The philosopher is constantly torn between two conflicting

desires: his desire for precision and his desire for comprehensibility. A good example of the philosopher's desire for precision is the following definition, from *The Dictionary of Philosophy,* of "explanation": "The method of showing, discursively, that the phenomenon, or group of phenomena, obeys a law, by means of causal relations or descriptive connections; or, briefly, the methodical analysis of a phenomenon for the purpose of stating its cause. The process of explanation suggests the real performation or potential presence of the consequent in the antecedent, so that the phenomenon considered may be evolved, developed, unrolled, out of its conditioning antecedent." This may be an ideal definition for some philosophers, but very few people will be able to understand it. On the other hand, the simple definition, "an explanation tells you *why* something happens," may be easily understandable, but it is so vague that it is completely worthless. It may be clear that the correct and fruitful definition must be somewhere between these two, but this goal is difficult to achieve.

Philosophy is associated in the minds of most of us with the usage of Big Words. We sometimes get a terrible feeling that the philosopher starts out with words like "transcendentalism," "neutral monism," "ethical relativism," "representative realism," "phenomenalism," and "a priori synthetic categories," and then builds up what he considers difficult ideas out of these simple ones. Now this picture may be exaggerated, but there is no doubt that philosophy books are filled with Big Words.

How do these come into use? It is fairly easy to reconstruct what must occur in the minds of people who construct these words. They write a large number of books and they find that there are long strings of plain old ordinary English words that occur again and again. It then becomes convenient to abbreviate these by new words, and these in turn occur in certain familiar combinations and again new abbreviations are introduced. These pyramid, and somewhere at the top of this pyramid a Big Word appears. This particular use of Big Words is very helpful. If it weren't for these, philosophy books would be even longer than they are now. As long as Big Words are introduced in such a way that at all times it is possible, at least in principle, to replace them by the strings of words for which they stand, there is no objection.

The difficulty arises, as in the case of vagueness, when Big Words are used to cover up ignorance. Sometimes they are introduced with a vague indication that they stand for a long string of words, with only hints as to what this string might be, and we are supposed to guess exactly what the author has in mind. This is the dangerous use of Big Words. A good example of this is a word like *"Gestalt."* It is a tremendously popular word with philosophers of a certain type, and yet most of them use it in as vague a way as possible. They claim it to be one of their fundamental concepts. They claim that with the one idea of *Gestalt* they can sum up a wide variety of facts which otherwise would lie disorganized. But somehow or other, we can search through their entire books and not find out exactly what the *Gestalt* is.

There seem to be two roads open to the philosopher. One road is to use an artificial language, such as Mathematical Logic; in this case the philosopher would follow the footsteps of the scientist. The other approach would be to try to reconstruct an ordinary language, such as English; basic English is an example of an attempt to do this. It may be worthwhile to take a careful look at this idea.

Instead of arguing about the relative merits of existing proposals for a basic English, let us start from scratch and discuss how such a language would be built up. We would begin with the small words in the English language which are clear to all of us. There may be few of these, but they would form a good starting point. We might call these the zero-level words. Built on this foundation, we could define words of the first level, all of whose definitions are given explicitly in terms of zero-level words. While we are doing this, we might find out that some of the words we would like to have on level one are so vague in ordinary usage that nothing but confusion would result from adopting them. Therefore, in this case, we would introduce entirely new words into the language. Using zero-level and first-level words, we could then define more complicated words, words of the second level, and so on, as long as we please. It might be worthwhile to measure how big a Big Word is by seeing at which level we finally reach it. In writing down these definitions, we would construct a dictionary of an entirely new kind.

Not only would our dictionary be free of vagueness and of ambiguity, but it would also be free of circularity. All the dictionaries of ordinary design are necessarily circular, since the author defines every word. The very words that he is defining occur in the definitions themselves. It is only by having zero-level words, which are left undefined, that we can avoid circularity. We could imagine that children are taught their words, starting with words of zero level, which are easy to understand, and increasing their knowledge, level by level, until even the biggest Big Words become as clear to them as the simplest word on the lowest level. This reconstruction of everyday language could overcome two of the great evils, vagueness and ambiguity, and thus make a major contribution to the elimination of useless verbal quibbles. But it is to be feared that even such an attempt would not overcome the difficulties arising out of emotional overtones of words.

This sort of program, while not as difficult as trying to adopt Mathematical Logic for Philosophy, is still a utopian program. Two major difficulties are evident. On the one hand, one has to construct such a basic language, which is a superhuman task. On the other hand, there is the quite natural hesitance on the part of people everywhere to accept such a new language. To mention but one of many difficulties in the way of acceptance, we must think of all the literature which would become obsolete and unreadable to a new generation that grew up speaking this language.

Although the goal of a perfectly precise language for Philosophy is utopian, there is nothing to stop us from trying to approximate this goal as best we can in our discussions.

This will be my guiding principle throughout this book. I will make every attempt to keep this book readable and yet to make my statements precise. I will try whenever possible to use a group of small words in place of a Big Word. There certainly seems no reason in principle why one cannot make the presentation of very difficult problems popular in the sense that many people can understand it without losing the essence of the nature of these problems. This has been illustrated in scientific writing. It is true that the original writings of most scientists are beyond the average intelligent reader, and the usual kind of popularization, though highly understandable, is full of mistakes. But a small number of excellent popular science books have combined the rigor of your creative scientist with the readability of an interesting novel. This is the goal I am aiming at in my discussion of Philosophy of Science. Of course, there are a substantial number of readers who judge how deep a book is by the number of Big Words occurring in it and by how difficult it is to understand various passages. To these readers I can only say that I hope they will find my book extremely shallow.

A VERY IMPORTANT WORD

There is one key word about whose meaning we must agree from the outset. This word, of course, is "Science." Unfortunately this word is ambiguous. There are two closely related but different meanings in which this word has been used. On the one hand, there is the German word, *"Wissenschaft,"* which applies to an extremely general type of intellectual activity, which has been translated as "Science" in English. On the other hand, there is the narrower usage of the word "Science" that is customary in this country. In the former usage of the word all types of scholarly activities, having very little relation to each other, are collected under a single heading. On the one extreme, we speak of a Science of Logic and a Science of Mathematics; at the other extreme, we speak of a Science of Ethics. If so many different things are all collected, it becomes extremely difficult to make any statements applicable to all of them.

It is for this reason that I want to reject the very broad usage of the word "Science." In this book, whenever I use "Science," I will mean a field like Physics or Chemistry, Biology or the Social Sciences, which are, or can become, scientific in the more common usage of this word. By restricting myself to this limited field I will attempt to show that there is a basic underlying activity common to all of these fields. I will try to show that, although there are many surface differences which are apparent on a first examination, fundamentally they all agree in the way in which they accept their theories. In the last analysis, in all of these fields the theories are led back to factual observations. They arise out of factual observations and they have their final justification in further factual observations. This is the common tie uniting various types of scientific activities which is not shared by other types of intellectual pursuits. I will argue that Science, in the sense in which I use it, has Mathematics as its tool, and, on the other hand, is used as a tool by Ethics and other disciplines. But I will try to argue that neither Mathematics nor Ethics is itself science.

These preliminary remarks enable us now to consider our first important problem, namely: How is Mathematics used in Science?

SUGGESTED READING

Complete references will be found in the Bibliography at the end of the book. More difficult selections are marked by an asterisk.

Basic English.
  Richards [1].

Vagueness.
  Hayakawa, Chapters 6, 11, 14.
  Beardsley, Chapters 2, 5.

Defense of vagueness.
  Richards [2], Chapter III.
  Benjamin, pp. 69-71.

Precise language.
  Mises, Part I.
  *Carnap [3].

2

Mathematics

"And if you take one from three hundred and sixty-five, what remains?"
"Three hundred and sixty-four, of course."
Humpty Dumpty looked doubtful. "I'd rather see that done on paper," he said.

SINCE IT TAKES some fifteen years of schooling to get an inkling of what Mathematics is all about, I am most skeptical about short accounts of this involved problem. Yet we must try to acquire a feeling for the nature of this subject. In high school we are usually handicapped by the fact that the teachers themselves do not understand the true nature of Mathematics. The student is bribed into taking an interest by promises that Mathematics will turn out to be useful in getting correct change or building pyramids or making atomic bombs. It is indeed an exceptional secondary school that will even attempt to teach what Mathematics is, as distinct from what it is used for.

Our only hope is to start from scratch.

WHAT MATHEMATICS IS NOT

First of all, this is a protest against all those high-school teachers who have tried to make their pupils believe that mathematical propositions are self-evident truths. Didn't it ever bother you that it took them weeks of hard work to convince you how self-evident it was that the base angles of an isosceles triangle are equal?

Let us take a rather famous mathematical proposition as an example: "Every even number can be written as the sum of two prime numbers" (a prime number being a number not evenly divisible by any number with the exception of 1 and itself). For example, $2 = 1 + 1$, $4 = 1 + 3$, $6 = 3 + 3$, $20 = 7 + 13$, etc. But is this true of *all* even numbers? Well, if mathematical propositions are self-evident, just look at it for awhile, and it should become evident whether this is always the case or not. Before you find yourself spending the rest of your life on this task, someone should warn you that the German mathematician, H. Goldbach, presented this proposition (as a guess) to the mathematical world quite some time ago and that the leading mathematicians have tried to prove or disprove it for years, entirely without success. In short, mathematical propositions are not self-evident.

Before we go on, it will be useful to consider just what kinds of propositions there are. For this purpose I must introduce four Big Words. One of the oldest questions about various problems is whether they can be answered by pure reason, or whether facts about the universe must be brought into the picture. We imagine the Greeks sitting in a secluded room contemplating and going outside to look at facts only as a last resort. Propositions whose truth (or falsity) can be shown by pure reason, prior to observations, are called *a priori* propositions. The remaining propositions, which can be decided only after facts are available, are called *a posteriori*. These two Latin phrases, derived from the words for 'before' and 'after,' are in such common use that I felt I must introduce them. There is a second common division of propositions which cuts (or was supposed to cut) across the previous one. A proposition like "All bald men are bald" gives you no factual information; its truth follows simply from the meaning of the words used. Similarly, the proposition "Some bachelors have beautiful wives" is false because of the way we use the word "bachelor." Such propositions, which simply analyze the meaning of some words, are called *analytic*. Most propositions, of course, do have factual content for example, the true proposition, "Some men are bald" and the false proposition "All men have beautiful wives."

Such factual propositions are called *synthetic.* We thus have four types of propositions according to these divisions.

	A priori	*A posteriori*
Analytic	Type 1	Type 2
Synthetic	Type 3	Type 4

You should note that this is a classification as to how the proposition can be decided and as to its content, but not as to truth or falsity. A proposition of any type could be either true or false, as seen by the above examples.

Let us consider analytic propositions first because they cause no trouble. They are the most certain propositions we have; we need only consider the meaning of the terms used to find out whether the proposition is true or false. Surely for this we need no factual evidence: pure reason will tell us that all bald men are bald. Thus we see that all analytic propositions are *a priori,* or that there are no propositions of type 2.

Synthetic propositions cause more trouble. Certainly there are many propositions of type 4; most ordinary propositions are factual, and we need to make observations in order to decide whether they are true or false. "All Americans are under 10 feet tall" expresses a matter of fact, and we must observe about 170 million people in order to check that it is true. The question is: Can thought alone tell us anything about the universe? Are there any such factual propositions which can be established by pure reason, i.e., are there any *a priori* synthetic (type 3) propositions?

The chief proponent in modern times of the view that there are such propositions was the famous German philosopher, Immanuel Kant. His theory is very difficult to explain and even more difficult to understand. But *very* roughly speaking, he held that we see the world through colored glasses, or rather that our minds impress a certain pattern on the physical world. This pattern is in terms of space and time and certain "categories," a word whose meaning we need not consider in this book. What interests us is that the propositions of Mathematics are, according to Kant, such type 3 propositions. They express factual propositions, since they tell us something about our experiences, but they can be known by pure reason, since it was the mind that stamped them upon reality. This is one of the most ingenious philosophical theories ever invented; it is really a pity that Kant was quite wrong. The reasons will become clear later in this chapter.

So we find that all synthetic propositions are *a posteriori* and that the only valid types are 1 and 4. On the one hand, there are propositions whose truth or falsity can be established by pure reason, but this is only because this follows from the meaning of the words used; on the other hand, there are the factual propositions which we can decide only on the basis of observations. To which of these two types does Mathematics belong?

The view that mathematical propositions are factual and hence that they rest upon observations was held by the great nineteenth century British philosopher, J. S. Mill, among others. He believed that the only reason we are certain of mathematical theorems is that they have been so well confirmed by observations that they are "practically certain." I will use two excellent examples constructed by C. G. Hempel to show why this view is not tenable.

If a proposition is factual, we carry out certain observations in order to check it; if the observations agree with the expectation, we accept the proposition; otherwise we reject it. But when we try to do this with mathematical propositions, we find that we simply do not behave in this manner. Suppose we want to test the proposition "$3 + 2 = 5$." Let us take three microbes and put them on a slide; then take two more and put these on the slide also. We count them. Suppose we find six microbes on the slide, not five. We might conclude that there was a microbe on the slide to start with, or that we miscounted, or that one of the microbes divided into two. But under no circumstances will we conclude that $3 + 2$ does not equal 5!

Again let us try to test that, if $a = b$ and $b = c$, then $a = c$. Let us take a shade of blue and another just like it, then take a third shade of blue just like the second one, and let us suppose there is a noticeable difference between the first and third shades. We still would not conclude that the mathematical proposition is false, but rather that there were imperceptible differences between the first and second, as well as between the second and third, shades of blue, but that these differences when totaled became noticeable. If we try to imagine other ways of testing the ordinary mathematical propositions, we find in each case that it is entirely irrelevant what the result of the observations is, we *never* reject the proposition. Therefore we must give up the view that mathematical propositions are factual.

There is no choice left. Mathematical propositions are analytic a priori. They consist of an analysis of the meaning of words. If that is all they are, why bother with them?

WHAT MATHEMATICS IS

Instead of answering this question let us ask a new question: What is the relation between Mathematics and Logic? Logic deals with the form of propositions and of arguments. This word "form" is a Big Word, even if it is deceptively simple sounding.

Of the various different meanings of "form," the one most helpful is the "form" you have to fill out. Think of an application for college or for a job, where you are given a form for biographical information and asked to fill in the blanks. Then, roughly speaking, they provided the form and you provided the content. Let us do the same for some propositions that we have considered before. Take "All bald men are bald." Its form is "All —— . . . are ——." (I have used two types of blanks to indicate that there are two words missing. The first and last blank are the same, showing that they must be filled in with the same word in each case.) From this form we can get not only "All bald men are bald" but also "All lazy hippopotamuses are lazy," "All dull books are dull," and an infinity of others. These statements all have the same form. We know that the first statement is true by virtue of its form; hence all the others are true also. In one stroke we have established the truth of an infinity of propositions, and this is the purpose of logic: by studying forms only, logic can find results with far-reaching applications.

Before giving more examples of forms, it will be convenient to introduce a new notation. The use of different kinds of blanks is very clumsy, and therefore mathematicians have invented a gimmick of using special letters, like *x, y, z,* as if they were blanks. They are called variables, because various sorts of things can be put in their place; in short, they are blanks to be filled in. Let us use this gimmick to study the form of a famous argument known as the *syllogism*. From "All cows are mammals" and "All mammals are animals" we infer that "All cows are animals." In other words, if the first two propositions are true, then the third one must also be true. This argument has the form: "If all *x* are *y* and all *y* are *z,* then all *x* are *z*." This is still a very simple argument, but it is typical enough to allow us to improve our picture of logic.

I have said that logic studies forms only, and I emphasized the importance of blanks. But you cannot build a form out of blanks alone. In our example we find the words "all," "are," "and," "if . . . then." These have a definite, fixed meaning no matter what we insert in place of the variables. I must now introduce the technical term "constant." A *constant* (in logic) is a word or other symbol which has a fixed, unchangeable meaning as contrasted with the variables. Specifically, the constants, like the four mentioned above, are called *logical* constants, since logic must use them to build forms. On the other hand, those constants which we use to insert for the variables are called *subject-matter* constants, since they supply the content or subject matter of the proposition.

These preliminary considerations enable us to tackle the nature of logic, and through this the nature of Mathematics. Logic studies the meaning of logical constants and the way they enter into forms of propositions. The advantage of concentrating on abstract forms is that the results found are applicable to *any* subject matter. It is important to learn that, although it is slightly more work to do things in a general way, in the long run a lot of effort is saved. If one general argument can take the place of infinitely many specific arguments, then general arguments are indispensable.

It may be interesting to mention one result of the study of logical constants. It can be shown that it suffices to have three constants at our disposal: "All," "neither . . . nor," and "are." All conceivable forms can be constructed in terms of these. Naturally, this result has vastly simplified the study of logical forms.

Thus we can sum up by saying that Logic considers the form of propositions. It starts out by considering statements. It studies the meaning of the words occurring and the rules of syntax according to which words are compounded. Finally it arrives at the abstract form of the proposition expressed by the statement. What Logic wants to know is whether a proposition of this form is necessarily true—or necessarily false. A quick look at the table of types of propositions will show that Logic is concerned with analytic propositions: it wants to be able to decide truth or falsity purely on the basis of form, without consulting facts.

You can now understand why logicians spend so much time studying the little building blocks of forms, the logical constants and the variables, as well as the mortar holding them together—the rules of syntax according to which statements are formed. By going to the very roots of languages, Logic comes up with results of the widest applicability.

So much for the nature of Logic—now to its relation to Mathematics. Bertrand Russell has eloquently summed up the relation as follows: "Mathematics and logic, historically speaking, have been entirely distinct studies. Mathematics has been connected with science, logic with Greek. But both have developed in modern times: logic has become more mathematical and mathematics has become more logical. The consequence is that it has now become wholly impossible to draw a line between the two; in fact, the two are one. They differ as boy and man: logic is the youth of mathematics and mathematics is the manhood of logic."

We must now see how the identity of the two fields can be demonstrated. This work was first done about the turn of the century. It can be shown that all of Mathematics is founded on properties of integers (whole numbers). If you are well acquainted with these, the rest of Mathematics is deducible by purely logical arguments. So in a sense the nature of Mathematics can be identified with that of the theory of integers. Late in the last century an Italian mathematician, Giuseppe Peano, tried to show that the properties of the integers follow from five simple postulates; hence all of Mathematics follows from these five propositions by pure Logic. Actually, this program was incomplete. But in the meantime modern Mathematical Logic—that is, Logic strengthened by the use of symbolic methods—was being developed by the English mathematician, George Boole, and the German logician, Gottlob Frege. Their work leads up to the all-important *Principia Mathematica* (a masterpiece written about 1910 that is discussed by practically every philosopher and read by practically none) by two of our greatest contemporary British philosophers, Bertrand Russell and A. N. Whitehead. In this work they show that the mathematical concepts used by Peano can be defined in terms of logical constants and that all their properties can be demonstrated by pure Logic. Thus Mathematics is shown to be no more than highly developed Logic. (In this process two new logical principles turn up, the axioms of infinity and choice, whose somewhat controversial nature need not concern us here. Let it suffice that if we recognize these two as legitimate logical principles—as most logicians do—then all of Mathematics follows and becomes just advanced Logic.)

Thus we see that Mathematics is a study of the form of arguments and that it is the most general branch of knowledge, but that it is entirely devoid of subject matter (since this is carried by the subject-matter constants, which are outside of Logic). We are forced to reaffirm that mathematical propositions are analytic and must ask again the question we raised at the end of the last section: What good are they? Only now we are in a position to answer the question.

As we have said, Logic (and hence Mathematics) as a study of forms tells us what follows from the meaning of the terms used. For example, in Humpty Dumpty's problem it tells us that $365 - 1 = 364$. The argument is somewhat as follows: "$- 1$" *means* going from a number to the previous one, "$=$" *means* being the same number, and hence the proposition asserts that the number 364 is the number right before 365; this is true, because that is how we *named* our numbers. Thus the proposition is true because of the meaning of the terms; it is a true analytic proposition. Your response to this is undoubtedly: "So what!" I grant you that this is hardly worth seeing on paper. But when you are told that it follows from the meaning of the words used that "every number can be written as a product of primes, and in a unique way" or that "the number of the primes up to n divided by n and multiplied by the natural logarithm of n tends to 1," you are less likely to scoff. It is not necessary that you should understand these famous theorems fully. My point is only that sometimes things follow from the meaning of words that are so far from obvious that it takes a good mathematician several weeks or even years of painstaking labor to prove them.

Here lies the secret of the success and of the vital importance of Mathematics. While in a sense Mathematics brings us nothing new, since it does no more than analyze the meaning of our words, it brings us facts that are *new to us,* facts we did not realize we possessed.

In a logical deduction, or mathematical proof, we start with a few propositions which for some reason we accept, and then we come up with entirely new propositions which we *must* also accept, because they are contained in what we have said before. And thanks to the complete generality of Logic (Mathematics), it can do this for any subject whatsoever.

THE RELATION OF MATHEMATICS TO SCIENCE

There is a branch of Mathematics which is ideal for a study of the uses of Mathematics in Science: Geometry.

Geometry dates back to the Egyptians, but in most of our minds it is connected with Euclid, the man who is credited with the first complete systematic treatment of the subject. Euclid laid down a set of axioms from which all true geometrical propositions followed by pure Logic. Actually we know today that Euclid's system fell short of its goal on two counts: First, it was redundant; it contained some definitions which served no purpose at all (this is a minor fault); and second, it made some hidden assumptions that did not follow from the axioms (which is very serious). But in this century the great German mathematician David Hilbert, corrected all faults and has given us a completely satisfactory system, which is still known as Euclidian Geometry. Hilbert's Plane Geometry has as undefined terms "point," "line," "is on the line," "lies between the points," "are at equal distances," and "are equal angles," which are used to formulate the axioms and to define all other terms; and it has a list of basic axioms.

Let us concentrate on the axioms for the moment. As you probably remember, an axiom is something taken for granted, something that is not proved. But if Mathematics is to do everything by pure Logic, why not prove the axioms too? The answer is that you cannot prove something from nothing, you must have something with which to start your first proof. This is quite analogous to the difficulty we ran into in the last chapter when we tried to define all words. To make up for the fact that these axioms had to be taken on faith, Euclid tried to choose them in such a way that they were obviously true, i.e., that they corresponded quite obviously to our everyday experience. Fortunately for Philosophy, he did not entirely succeed in this. There was the famous axiom about parallels which seemed much less evident than the rest. Instead of stating Euclid's own version, I shall give you the version you are more likely to be familiar with: "Given a line, and a point not on the line, there is just one line through the point parallel to the line." All the other axioms could be "checked" by making small diagrams, but this troublesome axiom stated that the two lines *never* meet, for which there is no practical test.

So for the next two thousand years mathematicians tried various remedies for this embarrassing situation. Many substitutes were tried for this axiom, but every adequate replacement proved just as difficult to test. Finally it was decided that the axiom isn't really an axiom, but a theorem. You see, there was the basic conviction that every true proposition of Geometry must follow from self-evident axioms, and if this particular proposition, though true, was not self-evident, then it must follow from the remaining axioms that are self-evident. To make a long story short, they did not succeed in proving the parallel axiom.

The story took a dramatic turn when around the beginning

of the eighteenth century, someone tried to use the indirect method of proof. The indirect method is an ingenious trick useful in all types of thinking: Assume the *opposite* of what you are trying to prove, show that this leads to a contradiction, and hence you prove your point. Take as an example the detective who tries to prove that a murder was committed by someone well acquainted with the victim. He argues as follows: "If the deceased did not know the murderer well, then the victim allowed a relative stranger to walk up to him, to pull out a gun, to put it to his temple, and let him pull the trigger, without putting up a struggle. But this is absurd. Hence the murderer must have been well known by the victim, well enough to take him completely by surprise." Quite analogously several mathematicians tried to argue as follows: "Let us assume that the parallel axiom does not follow from the others; then we will not get any contradiction if we deny it. Let us suppose that we can draw more than one parallel to a given line through a given point, and show that this leads to an absurdity." Note that these men were quite certain that the parallel axiom does follow from the rest, and hence they were disappointed when they failed to detect an absurdity in the modified system. They considered themselves utter failures, little realizing that they had one of the most thrilling discoveries of all time at their fingertips. This discovery had to wait until the 1820's.

As one of the strange coincidences in the history of human thought, after 2000 years of futile search, the solution was found by two men independently and at the same time. They were the Hungarian John Bolyai and the Russian N. I. Lobachevski. They tried for years to find an absurdity in the geometry with many parallels, and when they did not succeed they discovered a new kind of geometry, which was as good as Euclid's. (The proof of this fact was supplied later by Felix Klein.) Let us quickly add that their work was extended by two of the greatest mathematicians of all time, both German, C. F. Gauss and G. F. B. Riemann, who showed that there was a third basic type of geometry and that there were infinitely many "mixed" geometries. At this point it was high time that mathematicians stopped proving things for awhile and considered just what they were doing.

Let us for the moment return to Kant. Kant had the bad (or perhaps good) luck of dying some twenty years before the discovery of these new geometries. Had he lived longer, he would have had to abandon his views as to the nature of Mathematics and, hence, a good part of his philosophy. You may recall that he held that Geometry is certain, because it is imposed by our mind on the world. This, of course, presupposes that there is but one geometry; otherwise we are in the ridiculous position of having to choose from among three equally good systems and infinitely many mixtures of them, as to which is imposed on the world by our minds. This is an unanswerable objection. Kant had a most ingenious theory, and he can hardly be blamed for not being able to foresee future developments, but we must conclude that he was wrong.

The time has come for us to make use of what we have learned about the nature of Mathematics in the last section. We have said that Mathematics is no more than an analysis of the meaning of words, and a study of forms of arguments. Looking at the axioms of Geometry, we find that the usual logical constants (like "and") are used, and in addition we find the six subject-matter constants which we mentioned above (like "point"). But having stated that subject-matter constants are outside the realm of Mathematics, we are in the queer position of having to say that the words "point," "line," etc., do not belong to geometry. If we write "dog" for "point," "cat" for "line," "dog bites cat" for "point lies on line" and do this consistently, the theorems will follow just as logically as before. Of course, it may seem strange to have an axiom stating that for every pair of dogs there is a cat which they both bite, but you will get used to this after awhile.

Before you are completely bewildered, let me state quickly that this is not what mathematicians do. As we saw before, Mathematics eliminates subject-matter constants by replacing them by blanks, or variables. Thus we put x for "point," y for "line," R for the relation between a point and a line "is on the line," etc.

Thus instead of saying that "a point lies on a line" or saying that "a dog bites a cat," a pure mathematician would say that "an x bears the relation *R* to a *y* or more briefly that "*x*Ry." This is done for all the axioms and all the theorems, until we have a "skeleton" of them. And then we show that the theorems still follow from the axioms, by nature of their form. In this it is most significant that we make no use of what *x, y, R,* etc. mean—as a matter of fact they do not mean anything until we give them a meaning. Thus, *no matter what* they mean, the theorems will follow from the axioms. Therefore, *if* we can find any meaning at all for these six entities which will make the axioms true, *then* the theorems will be true under the same interpretation. This is the "if—then" nature of Mathematics. This shows both its strength and its weakness. It can prove perfectly general statements true about all possible applications; on the other hand, it cannot prove the truth of any one statement in an application—this is left to Science. All that Geometry tells the scientist is: "Find me *x*'s, *y*'s, *R*'s, etc., which make these axioms true, and I will give you as a present an infinity of true statements—all my theorems."

THE THREE GEOMETRIES

First systematic treatment by	Bolyai and Lobachevski	Euclid	Riemann
Sum of the angles of a triangle	Less than 180°	180°	More than 180°
The space curves	"Out"	Doesn't curve	"In"
Volumes increase	Rapidly	"Normally," i.e., like cube of radius	Slowly
To a given line, through a point the no. of parallels	Is infinite	Is 1	Is 0
Such a Universe would be	Infinite	Infinite	Finite
Behaves like a	Saddle	Plane	Sphere

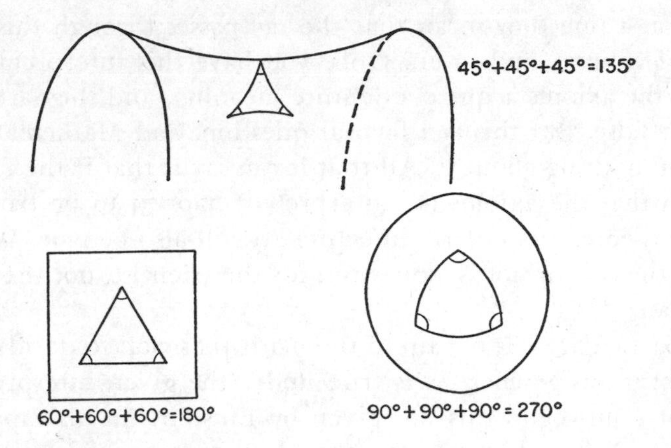

An immediate consequence of this is that Mathematics can tell you nothing about the truth of your axioms; as a matter of fact it is nonsensical to ask whether the axioms are true! Would you care to state whether the axiom "Given any two x's there is a y such that both x's hold the relation R to the y" is true? The only sensible answer is: "It all depends on what x, y, and R are." If we have our dog-cat-bite interpretation, then it is clearly false; it is not true that for any two dogs there is a cat they both bite. But if the x's are men, the y's women, and R the relation of liking, then it seems to be true that for any two men you can find a woman whom they would both like.

We find that statements of Mathematics are only forms and that their truth can be considered only if we fill in the blanks, if we give them an *interpretation*. This is the distinction between pure and applied mathematics, which will occupy us in the next section.

I will now raise the question which has, I hope, bothered you for some time: "How can you say that the geometrical axioms are only forms, that they are neither true nor false, when Science makes such frequent use of them?" The answer is that when scientists use these propositions, they automatically fill in the blanks; they have very definite meanings in mind for each of the six terms. By a point they mean the smallest volume they can identify, by a line the path of a ray of light in vacuum, by a point lying on a line they mean that the ray passes through this infinitesimal volume, and so on. Once you have this interpretation in mind, the axioms acquire a definite meaning, and they are either true or false. But this is a factual question, and Mathematics can tell you nothing about it. All that it can say is that if the universe is such that the axioms (so interpreted) happen to be true, then all the theorems (similarly interpreted) will also be true. Whether this is the case or not is a question for the scientist, not the mathematician.

In particular, it is not up to the mathematician to decide which of the various geometries is true under the given interpretation. The first answer to this was given by Einstein (in his capacity as a physicist), and this answer is that the universe is apparently *not* Euclidian. It is a mixture of these three basic types, varying from place to place, according to the distribution of matter in it. As to the universe as a whole, we are not sure, but it seems to be finite, though endless, curving back upon itself, which no self-respecting euclidian universe would ever do.

I will not go into this deeper; this is a scientific question, the various proposed answers to which you can find in any good popular book about the universe. Let me instead quote Einstein's famous statement: "As far as the laws of mathematics refer to reality, they are not certain, and as far as they are certain, they do not refer to reality." Take the second half of this statement first: The laws of Mathematics are certain in their formal, analytic status. In this they do not contain any subject matter, and hence do not refer to reality. If, however, we interpret the axioms, then they refer to reality, but they are no longer mathematical statements and we can never be certain of them. The certain statement is that the theorems follow from the axioms, but this is language-analysis, not Science; the axioms themselves, when interpreted, are scientific, but are not certain in themselves.

Then why is Mathematics so vital for Science? It is an ideal language in which to formulate our scientific theories, and it tells us just what our theories imply. Contrary to popular belief, we rarely (if ever) test a scientific theory directly; what we test are logical, that is, mathematical, consequences. Consider Geometry as interpreted above, and take in particular the parallel axiom. There seems to be no practical method for testing this directly. However, one of the consequences of this axiom together with the other axioms is that the sum of the angles of a triangle is 180 degrees. In the other two types of geometry we get sums of more or less than 180 degrees. This can certainly be tested. Gauss actually did this, but unfortunately the result was not conclusive. You see the catch is that for small triangles, with sides of no more than a few billion miles, the difference predicted by the three theories is negligible, and our measurements are not accurate enough to tell the difference. This, incidentally, is the reason why Euclidian Geometry is all right for everyday use! But some day we may have more accurate instruments, or be able to measure huge celestial triangles, and then we may find that the sum is different from 180 degrees for some triangles, and indeed this is what Einstein leads us to expect.

A more useful way of differentiating among the geometries is the formula for volumes. You may remember that the volume, according to Euclid, increases as the cube (third power) of the distance. In the geometry that curves in on itself this happens more slowly, whereas in the third geometry, volumes increase more rapidly. Our present observations indicate that the number of the nebulae (large groups of stars) increases more slowly than the cube of the distance, thus—assuming that they are evenly distributed, which we do assume—it seems that the Universe curves in. But very recent data casts doubt on this conclusion.

To sum up: Mathematics is invaluable for Science because, by showing us an endless number of statements which are contained in our theories, it gives us an infinity of ways of testing our theories. We will return to this whole problem in Chapter 5.

PURE VS. APPLIED MATHEMATICS

Let us now make up a mathematical system. (This is really much easier than it sounds.) First, we will state it in its pure form, with variables, and then immediately—to make it easier to follow—we will supply subject matter to the axioms.

AXIOM 1. There is an x such that no other x bears the relation R to it.
AXIOM 2. No x has more than one x bearing the relation R to it.
AXIOM 3. Each x bears the relation R to just one other x, and never bears this relation to itself.

Let us now interpret the x's as dogs and the relation R as biting.

AXIOM 1′. There is a dog who is not bitten.
AXIOM 2′. No dog gets bitten by more than one dog.
AXIOM 3′. Each dog bites one other dog, but never bites itself.

It is pretty clear that these axioms are not all true about our world. Certainly the second axiom is false, and presumably so is the first part of the third. But the mathematician does not care about this. He just says that *if* these axioms are true, *then* certain other facts also hold. Let us derive one such consequence. I will use the interpreted form, to make it easier to follow.

THEOREM: There are infinitely many dogs.
PROOF: I will use the indirect method. Let us assume that there are but a finite number, and get a contradiction from it. From Axiom 1′, we know that there is a dog not bitten, and we will call him the first dog. From Axiom 3′, we know that he bites another dog; call this the second dog. This dog in turn bites a dog; call him the third dog. And so on. But if there are only a finite number of dogs, we must come to an end. Consider the *last* dog in this list. Whom does he bite? He must bite some dog (Axiom 3′), and he can't bite himself (Axiom 3′ second part), so it must be an earlier dog. But it cannot be the first dog (Axiom 1′), and all the other dogs on the list got bitten by the dog just before them, and hence they cannot be bitten by the last dog (Axiom 2′). Thus there is no dog for the poor last dog to bite, which contradicts our third axiom. Hence we must have been wrong in supposing that there were but a finite number of dogs, which proves our theorem.

This, strangely enough, is a perfectly rigorous mathematical proof. We have established that our three axioms guarantee the existence of infinitely many x's (dogs). This is an excellent example of a theorem which contains nothing new, since no mathematical theorem adds anything to the axioms, but which still comes as a considerable surprise to us. It is most useful to know that this is implied by our axioms: If we know that there are only a finite number of dogs in the world, we can and must reject some of the axioms.

But does this make the axioms useless? No. There could be entirely different interpretations which make the axioms true. I will now give one such interpretation. Let the *x*'s refer to the years A.D., and the relation *R* be that of coming immediately before. Our axioms now become:

AXIOM 1″. There is a year A.D. which is not preceded by any other year A.D.
AXIOM 2″. No year A.D. is preceded by more than one such year.
AXIOM 3″. Each year A.D. precedes one other such year, but never precedes itself.

These are clearly true. The first axiom states the existence of the year 1 A.D. The second states that there is always at most one previous year, and the third that there is always a next year. Since all the axioms so interpreted are true, all theorems so interpreted must be true. Indeed, our theorem now states that time is endless, which is true.

In this simple example we can see very plainly the difference between pure and applied mathematics. Pure mathematics states that any entities *x* and relation *R* which happen to satisfy the three axioms must also satisfy the theorems. It says nothing about what *x* and *R* are, and nothing about the truth of axioms (which are so far only forms and hence cannot be said to be true or false). We pass to applied mathematics by interpreting the variables, by saying what *x* and *R* are. Then it becomes a matter of fact whether the axioms are true, and the problem is taken out of the hands of pure mathematics and put in the hands of Science. Applied mathematics belongs to Science, and I will argue that all of Science is applied mathematics. This, I believe, is the reason why Mathematics is often classified as a science. It also provides the final answer as to the status of mathematical propositions.

Propositions from pure mathematics are analytic a priori, whereas applied mathematical propositions belong to Science, and thus are synthetic a posteriori.

In order to avoid the customary confusion, I will always mean pure mathematics by the word "Mathematics," unless I explicitly qualify it.

There is another way of stating this difference, which may clarify the situation. The x's and R's are undefined in Mathematics. They can be used in turn to define other words, but they form the basic vocabulary which is not defined. Thus, in Geometry, we define angles in terms of lines, etc., but lines are never defined. I should say that they are never defined explicitly. We never say that lines are so and so, because (as we saw in Chapter 1) some words must remain undefined. But in a sense they are at least partially defined. We do state that basic terms satisfy the axioms, and this characterizes them somewhat. We do not say in the above example what x and R are, but we do say that they are the kind of things which have the properties stated in the three axioms. This is called an *implicit* definition. Now put yourself in the place of the scientist. Suppose that he finds some objects and a relation which do have these properties; then he can interpret the x's to be these objects and R to be this relation (for example, years A.D. and coming just before), and Mathematics will tell him immediately that his objects and his relation have all the properties ascribed to them in the theorems. For example, he will know that there are infinitely many of the objects in question.

Now let us turn the problem around. Suppose the scientist has some entities he wants to study, together with certain relations holding between them. He will then search for a branch of Mathematics whose system of axioms, when interpreted, will correctly describe his entities and the relations he wants to study. Finding such a mathematical system and interpreting it to suit his purposes are known as "forming a theory." This is why it is of vital importance that mathematicians carry on their research without regard to applications. They can never tell which of their purely formal theories will someday be usefully interpreted by the scientist and become vital for mankind. The new geometries are an excellent example of this. Even after it was recognized that they are "as good as Euclid's," they were still regarded as intellectual curiosities without practical applications. Less than a century after their first development, and only a handful of years after mathematicians really learned how to work with them, they were given a physical interpretation, and now are central to the General Theory of Relativity.

One of the most difficult questions raised in this chapter is the manner in which mathematical propositions are interpreted. I have given some examples, but so far have not attempted to give general rules. Roughly speaking, a scientist takes a mathematical proposition and substitutes for the variables certain constants, expressing concepts which can be directly observed or measured. In this manner he takes the mere forms and, by filling in the blanks, he turns them into statements of facts that he can check. For example, interpreting the geometrical axiom that two points determine a line by the above interpretation, we get the factual statement that there is just one possible light path through any two points in space. That is open to scientific experimentation. This problem will be considered in detail when we come to Chapter 7 on scientific concepts.

CAN ALL SCIENCES USE MATHEMATICS?

The answer is "Yes." What is more, they *must* use Mathematics.

But you will often find the claim that the physical sciences are mathematical and the Social Sciences are nonmathematical. The reason for this misunderstanding is that people associate Mathematics with numbers. While I am quite certain that numbers will play a fundamental role in all these sciences soon, I want to maintain more, namely, that all scientific theories—numerical or other—are mathematical. This fact rests on the nature of Mathematics, on its identity with advanced logic.

When a scientist states a theory precisely and is interested in knowing just exactly what his theory involves, he is practicing mathematics. Take his theory, put blanks (variables) for his subject-matter constants, consider just what these forms imply, and you have a branch of Mathematics. This result is quite true in general, but some people (even scientists) are not convinced by general arguments; so let us take an "obviously" nonmathematical theory. There is a biological theory that the development of the individual in his early stages mirrors the development of the race. For example, the human embryo goes through various stages corresponding to the stages by which the human race developed. (This is why in the early stages the human embryo is practically indistinguishable from the embryos of lower animals.) Surely this is completely unmathematical!

Well, let us see. If we stated this theory quite precisely, we would have to arrange the development of the embryo in stages, and also the development of the race. We would get two series of stages. There are in addition certain properties which we study. Then we state that we can let the stages in one series correspond to the stages in the other in such a way that wherever one of these properties shows up in the embryo, the same property showed up in the race at the corresponding stage. In abstract form we have x's (stages of the embryo), y's (stages of the race), a relation of correspondence between them, say R, and certain properties. We would first of all have to state that the x's and y's form two series. This would give us, for each series, axioms like the three used in the example. Then we would have to state that R is a correspondence, the axioms for which are well known in Mathematics, and that the properties belong always to corresponding stages. This is what the mathematician means when he says that he has two isomorphic series (isomorphic with respect to the given properties). The study of these isomorphisms belongs to Topology, one of the many branches of Mathematics not using numbers.

We conclude that any precisely stated theory which is investigated as to its full implications becomes a branch of applied mathematics. In the foregoing example we find—what is so often the case—that the branch of Mathematics to which the scientist is led by his research has already been developed in its pure, abstract form by the mathematician.

The only reason some scientists deny that they use Mathematics is that for elementary theories the amount of mathematics needed is so little that the scientist can do it by logical intuition. But the difference between this and the use of the most powerful tools of modern Mathematics is only one of degree. As each science develops, it is forced to acknowledge its indebtedness to Mathematics more and more.

Thus we find that this irreplaceable language of Science, Mathematics, can never supply anything new. Yet it must be used constantly, so that we realize fully what we already have; and it never ceases to amaze us in showing that our innocent-sounding statements have consequences far beyond what we ever dreamed was contained in them.

SUGGESTED READING

Complete references will be found in the Bibliography at the end of the book.

Nature of Mathematics.
- Cohen and Nagel.
- Hempel [3].
- Wilder.
- Whitehead.
- *Carnap [1].

Mathematics and Science.
- Hempel [2].
- *Carnap [4].

Non-Euclidean Geometry.
- Courant and Robbins.
- Robertson.
- Poincaré.
- Gamow [2].

Prime numbers.
- Courant and Robbins.

3

Assumptions

> *"Always speak the truth—think before you speak—and write it down afterwards."*
> *"I'm sure I didn't mean—" Alice was beginning, but the Red Queen interrupted her impatiently.*
> *"That's just what I complain of! You* should *have meant! What do you suppose is the use of a child without any meaning?"*

THERE ARE SOME fundamental questions a scientist must ask himself before he starts his work. How can he justify his attempts as a scientist? What must he assume about the universe in order to enable him to hope for some kind of success? As a scientist he is going to be called upon to predict the future. He has seen only the past and only a small part of that. What right does he have to assume that the past is a good guide for the future? It is these questions that we must try to answer in the present chapter.

THE LAW OF NATURE

The usual argument proceeds as follows: A scientist searches for laws of nature. His search is hopeless if there are no such laws. Therefore the basic assumption of all Science is that there are laws of nature. This sounds very convincing indeed, but we have been forewarned about making sweeping assumptions. When we state something exceedingly general, the danger is that we may be saying nothing at all. My first major point will be precisely this, that the so-called assumption in this very broad form is empty of factual content.

Of course, no one denies that scientists have formed certain theories which have been referred to as laws of nature. We may be tempted to say that since scientists have found such laws, therefore laws exist. But we know that many of these so-called laws have turned out to be only approximately correct or, indeed, incorrect in many cases. Even a theory so well confirmed as Newton's laws turned out, in the long run, not to hold. Therefore the problem is not that of finding theories which hold approximately in limited range, and of which we are fairly sure, but the problem of laws holding without exceptions, exactly, throughout all time. We may suspect that some of the theories now held are indeed of this kind, but a suspicion is not enough. The question is: Can we really assume that such universal laws exist?

What is a law of nature? It is a general statement about the universe which is true. We know that these statements or theories are mathematical propositions which have been interpreted correctly. Thus, our question amounts to asking whether there are any general mathematical propositions which can be so interpreted that they are true? It is my contention that, if we formulate our assumption in this broad sense, we can prove without reference to factual information that there are indeed laws of nature. In other words, I will maintain that the so-called assumption is true, but analytically true.

It is a great pity for the Philosophy of Science that the word "law" was ever introduced. The usage of "law" carries the connotation that it can somehow be disobeyed. The idea of obeying or disobeying should not have entered these discussions at all. I am afraid this is part of our tendency to try to recreate the universe in our own image. It may perhaps date back to the primitive idea of God. To primitive people, God is but a very large and powerful human being. We even find examples where they try to make deals with Him. This may be in the form of sacrifices or promising to be good if he will only do them a certain favor. In mythology we find many examples where the gods are endowed with human strength and human weaknesses. It is perhaps a remnant from these days that we associate the idea of a law with nature. Presumably the laws of nature were passed by God. They were laid down on a heavenly piece of paper and these are the things that we must do. These are the things that nature must obey. There are even recorded examples of violations of the laws of nature, and these are called miracles.

We must realize that nature is not like a human being. Nature cannot obey or disobey. The laws of nature do not prescribe but, rather, they describe what happens. A law of nature is but a description of what actually takes place. In contrast to this, whenever a human law is passed, then it must be possible to violate this law. When we pass a law forbidding murder it is, of course, assumed that it is humanly possible to commit murder. If it were impossible to commit murder, then there would be no point at all in passing a law against it. Therefore we admit the possibility of someone acting against the law and we have to invoke penalties and enforcement agencies to see that the law is obeyed. But a law of nature cannot be disobeyed. There is no need for any kind of enforcement agency. The reason it cannot be disobeyed is much simpler. Since a law of nature is no more and no less than a careful record of what actually happens, there is no possible way of violating it. If in any experience one keeps a careful record of all that actually happens, then in no sense can this record be violated. After all, if we have stated exactly what has happened, then nothing can ever be done to undo this.

As a matter of fact, from this point of view a law of nature is a very dull thing. It is like a court record of proceedings on an average hearing before the judge. The only thing that lends it glamour is the fact that it is a record not only of the past, but also of the present, and most important of all, the future. It is like a court record of all past, present, and future proceedings. Anyone who is in possession of such a record will know what lies ahead of him.

Some years ago Hollywood produced an interesting movie on this subject. It dealt with a man who is given a newspaper each day. But by some unexplained method this paper turns out to be tomorrow's newspaper. Therefore, each day he knows what tomorrow's headlines will be. Naturally this gives him tremendous power over his fellow human beings. He can go to horse races and win any amount of money that he pleases. He can use the information available to him to get a good position and to advance himself and his friends. Of course, there has to come a moment in a Hollywood movie when the hero is punished for all this undeserved wealth. One day he reads that tomorrow's headline will be a report of his own murder. He does everything possible not to get near the place where the murder is supposed to take place, but he does not succeed. Through a strange series of circumstances he ends up in just that building and at just the right time. A shooting takes place and he is accidently shot though not killed. But he is reported dead, and indeed the report of his death is tomorrow's headline.

The laws of nature are, in a sense, like tomorrow's newspaper, but they report not only what happens the next day, but a week from now, a year from now, or many millions of years from now. In addition, the laws of nature are like the perfect newspaper. They never report anything mistakenly and they always include every last detail in their report. Any man in possession of a law of nature would have the most tremendous power ever dreamed of by mankind. It is for this reason that scientists spend their entire lives trying to get a small part of tomorrow's newspaper.

We are now in a position to prove the existence of laws of nature. Actually we can prove something stronger than that. We can prove that any given phase of the world is covered by some mathematical law correctly interpreted. Let us select an example that is supposed to be particularly unpredictable. Naturally I mean the weather. To make the example more specific, let us take the temperature of New York City. First of all, we must select a suitable mathematical language. But this has already been done for us. We will state our law in degrees Fahrenheit, the usual measure of temperature. What form would the law have to take? It is supposed to state for any moment in the past, present, or future just what the temperature was at the place that is now New York City. Such a correspondence is known in mathematical terminology as a function. More specifically, we have to describe the degrees Fahrenheit in New York City as a function of time. This is symbolized by writing $f(t) = d$. The question then is: Is there any mathematical function which, interpreted in the way we have just indicated, correctly describes the temperature in New York City at various times? But as soon as we have stated the problem this clearly it is very easy to see that such a function exists. Given any moment of time, there is no difficulty, in principle, in recording the temperature in New York City. Again, in principle, there is no difficulty recording this for all moments, past, present, or future. And this record defines a function. Of course, in practice we could never carry this out, but this is relevant only to the question as to whether we can find out what this function is, not to the question as to whether such a function exists. The possibility of making such a record, in principle, is sufficient to guarantee that this is a well-defined mathematical function. Hence it proves that this aspect of the world is covered by a law of nature. A similar argument will work for any other phase of the entire universe.

Once we have seen that the existence of a law of nature is no more than the theoretical possibility of keeping a careful record, we can do much more than we have done so far. It is possible, in principle, to keep a record not only of one event, but of any event in the entire history of the universe. If we are then prepared to imagine that some all-powerful heavenly agent keeps a careful record of all events in the universe, then these records, together, would form a law which covers everything that happens in the universe. Let us call this record *The Law of Nature.*

Naturally it is not claimed that there is only one possible way of constructing The Law of Nature. Since this law is not a pure mathematical law, but an interpreted one, there are many different ways of connecting mathematics with the universe. For any choice of an interpretation we will have one version of The Law of Nature. But since any one version is as good as any other one, we will suppose, for the sake of the argument, that one definite version has been selected and that this is our standard reference book in terms of which we can discuss the basic assumptions of Science.

We have now reached the surprising conclusion that there is no need to assume anything about the existence of the laws of nature. Then why have so many eminent people stressed the necessity of making such an assumption? The only reasonable conclusion is that, although the assumption usually found in books is analytically true, there must be a slightly different version of this assumption which makes a good deal of sense. Indeed we will find that this is the case.

THE NEED FOR ASSUMPTIONS

Let us try to get some impression of the form of The Law of Nature. It is a record of everything that happened in all parts of the universe in the past and the present, and everything that will happen in the future. Naturally, this is unimaginably complex. Even if we restrict ourselves to the temperature in New York City, we may find that the record is beyond the abilities of human bookkeeping. We can record the temperature for any given day; we can even record it for any given hour or, perhaps, for one-minute intervals. We might even be able to imagine that at some future time we could record it for every billionth or billionth of a billionth of a second, but we cannot possibly get it for every moment. Indeed, if someone presented us with such an infinitely long record, it is very doubtful that we would be able to make any use of it. This gives a clue as to what the true assumption must be. It presumably deals not with what nature is, but with our ability to comprehend nature.

This argument can be made clearer if we investigate one of the most fascinating branches of modern mathematics, the study of the infinite. After thousands of years of confusion in this field, Georg Cantor has enabled us to discuss infinite quantities with perfect confidence. We used to think simply that anything that had no end was infinite and that was that. Today we know how to differentiate between various orders of infinity. What might be said is that one thing is more infinite than another thing. More accurately, one quantity is of a larger order of infinity than another. Although this seems puzzling at first, it is really not very difficult to explain. The key question is: How do we compare infinite quantities?

Let us instead ask ourselves just exactly how we compare finite quantities. Our first answer might be that we count them and thus we are able to compare them. While this is perfectly acceptable, there is often a much quicker way of getting the answer. Let us suppose that we have a number of children in our house and we give them a basket of apples. Each child eats one apple and then all the apples are gone. We reach the natural conclusion that there were just as many apples as children. There was no need to count either the apples or the children. Or again let us suppose that students walk into a lecture room and take seats. If we find that all the students are seated and yet there are empty seats left over, we will know that there were more seats than students. If we find that every seat has been taken and there are students standing in the aisles, then we will reach the opposite conclusion. In neither case was it necessary to count either the number of students or the number of seats. This simple process of matching is our basic tool in comparing quantities. Although only paradoxes result if we try to apply our ordinary notions of counting to infinite quantities, we have no difficulty at all in matching them. Given two infinite collections, if we can match them up they are of the same order of infinity. If whenever we try to match them there are always things left over in the second collection, we reach the natural conclusion that the second collection is, in some sense, larger than the first one.

Thanks to the work of Cantor, we now have not only this basic tool, but far-reaching results in the field of infinite quantities. We will discuss some of the results in a highly simplified form. For example, we will ignore the fact that some important questions concerning the orders of infinity have not as yet been decided. However, the solution of these unsolved problems would either agree with the following presentation or make the differences even more extreme.

One well-known infinite collection is the collection of integers, 1, 2, 3, etc. There is no question about this collection being infinite since it starts, goes on, and never comes to an end.

Cantor has shown that this is the smallest infinite collection. There are other collections that are exactly as large as this, but there are none smaller. On the other hand, there are many other very large infinite collections. As a matter of fact, for any infinite collection that one can conceive, it is possible to find an even larger infinite collection. And thus we find ourselves with an infinite number of larger and larger infinite collections. Fortunately, we will have nothing to do with anything beyond the first three orders of infinity. We have said that the integers are of the first order of infinity. We will a little later be confronted with the second order of infinity. At the moment we are concerned with the third order of infinity. When we try to count up all possible laws of nature in the sense defined above (that is, all real-valued functions of a real variable), we find that this collection is of the third order of infinity. This gives us a major clue to our difficulties in understanding the laws of nature.

Let us compare this tremendous order of infinity with our human abilities to comprehend. In any human language we can make a list of all words, and hence of all sentences that can possibly be formed in the language. While there seemingly is no end to these, there is also no difficulty numbering them 1, 2, 3, etc., in some lexicographic order. Thus we can see that we can match sentences in a human language against the integers. This can also be done for all human languages, past, present or future. Therefore, we are forced to reach the conclusion that all laws comprehensible to human beings are a collection of the lowest order of infinity. As contrasted with this we find that all possible laws of nature are two orders of infinity higher, namely, of the third order of infinity. There is a tremendous gap between what we limited human beings can understand and what possibilities there are for nature. It would indeed be very surprising if The Law of Nature were one of the small number of laws that we finite human beings can understand. Now it is clear what the assumption for scientists must be: they must assume that The Law of Nature is sufficiently simple that we human beings can express it in our limited language.

This brings us to the end of a rather long and trying search.

We found, surprisingly enough, that there is no assumption needed to assure us that nature "obeys" certain laws. Since a law is no more than a description of what actually happens, then by the very meaning of the words it follows that there are laws of nature. We have even found that there must exist a single law applying to all of nature. On the other hand, we find that most laws that could conceivably be created for the universe will forever lie beyond our limited human possibilities. Hence in a sense the odds are infinitely great against our being able to comprehend The Law of Nature. Thus, if we shift our assumption from one about nature to an assumption about our human capabilities, we get a very strong assumption indeed. We must assume the very unlikely fact that The Law of Nature, or at least some laws of nature, fall into the limited range open to human beings.

VARIOUS TYPES OF ASSUMPTIONS

So far we have only asked in general terms whether it is possible for human beings to learn The Law of Nature or individual laws of nature. Actually there are three different types of assumptions that might come under this heading: (1) We may assume that it is possible to learn a law exactly; (2) we may assume that we can learn a good approximation to a law; or (3) we may assume that we can approximate a law as closely as we wish. Let us consider each of these cases briefly.

For practical purposes we never have to know a law exactly. We will always be satisfied if we have a sufficiently good approximation to it. If we are interested in the weather in New York City, to know it within 1 degree Fahrenheit will suffice for most purposes. If we know it within 1/100 of 1 degree, this will satisfy nearly every scientific need. And certainly we will never have to know the temperature more accurately than to 1/1,000,000 of a degree. We will all agree that in some sense a sufficiently good approximation to a law is as good as knowing the law exactly. But just what is a sufficiently good approximation? As a matter of fact what would we be willing to accept as an approximation to a law? Suppose that we try to get an approximation to the law governing the temperature of New York City. Would it be sufficiently good approximation to state that this temperature is always 50 degrees Fahrenheit? While it is intuitively clear that this is not acceptable, nevertheless we would be within 40 degrees 90 per cent of the time, and 75 per cent of the time we would be *much* closer. With just slightly more ingenuity, say by guessing a different temperature for each of the four seasons, we could get much closer to the actual answer, without in any sense coming near to the law governing temperature. Naturally we all feel that we have missed the point somehow. But while intuitively this may be clear, it is very difficult to see how one would make this requirement precise. Therefore, assumptions of type two still need a great deal of research. This is a very fruitful field in which a philosopher of science might make a contribution.

In the previous section we considered only the possibility of finding a law exactly. We have now mentioned the second possibility of finding a good approximation to a law. The third type of assumption would cover a situation somewhere between the first two. While we do not find out a law exactly, we find an approximation to it and then a better approximation and increasingly better approximations, getting as close to the exact law as we wish. This kind of assumption is probably the one that is closest to what the scientist actually has in mind. For this reason it is worth taking considerable trouble to explain. But considering approximations to a law is much too complex mathematically, and a simpler example will do as well. Let us consider approximating a given number. The example we will choose will be the number *the square root of 2.*

This number has a most intriguing history. For the story we must go back to the notable Greek mathematician and philosopher, Pythagoras, who lived about five hundred years before Christ. Although he predates modern Science by more than two millennia, he seems to have been the first man to appreciate the true significance of Mathematics for an understanding of the universe. Naturally, most of his teachings mixed a good deal of sound theory with a very large amount of mysticism. For example, he assigned mystical significance to various individual numbers. In addition to that, the mathematics of which he had command was highly limited. As far as numbers were concerned, he dealt primarily with the integers. He added them, subtracted them, multiplied them, and divided them. With these limited means, the most general numbers he was able to express were numbers that were ratios of two integers. These ratios were very successful in many fields, notably in music. Hence, the Pythagorians built an entire philosophy around these and were firmly convinced that all the secrets of the universe could be expressed in such simple ratios of integers. It can easily be understood what a shock it was to them to find that there are quite clear-cut exceptions to this conjecture. What was even more shocking was the fact that the exception was found in the well-beloved part of Mathematics, namely Geometry.

Consider a square whose side is of unit length. We want to find the length of a diagonal of this square. Since the diagonal and two sides form a right triangle, by the theorem bearing the name of Pythagoras we know that the square of the diagonal is equal to $1^2 + 1^2 = 2$. Hence the length of the diagonal is the square root of 2. This was well known for a long time. The shock came when one of the Pythagorian disciples succeeded in showing that this number cannot be expressed as the ratio of two integers. The disciples were tremendously shocked and thought that their entire theory about the significance of Mathematics crumbled. Actually it has since turned out that they could easily have overcome this difficulty by accepting a more general definition of what a number is. But we are told that their horror was so great that the discoverer of the exception to the theory committed suicide.

Consider the problem posed for the Pythagorians by the square root of two. If in their mathematical language they were only allowed to express those numbers which are ratios of two integers, then the square root of two could not be expressed in their language. This is analogous to the situation in which The Law of Nature is not expressible in a human language. But certainly Pythagoras knew many good approximations to the square root of two. Since the square root of two can be written as 1.4142 . . . , we can certainly call it, in first approximation, 1. This approximation is off by nearly one half. If we write 1.4 or, to put it in the form of a ratio 14/10, we will be off by less than 2/100. If we write 1414/1000, then we will be off by about 2/10,000. By continuing this method of approximation we can come as close to the square root of two as we desire, though we can never reach it exactly. Hence, in the language available to Pythagoras, there exist approximations as close to the square root of two as desired, though the square root of two is not in his language.

Thus we arrive at an assumption of the third type. We may assume that The Law of Nature (or certain individual laws) are not expressible in any human language, but that it is possible to find approximations to them as closely as desired. If this is the assumption that we make, then we get an extremely attractive picture for the role that Science plays. Science has set itself an all-inspiring goal, namely to learn The Law of Nature. This is a goal that Science can never reach. But it can come closer and closer without end. It is a picture holding out the hope of eternal progress without the danger that succeeding generations will find nothing worthwhile to do.

CAUSAL LAWS

Let us return to our original problem of trying to find factual content for the empty assumption about the existence of laws. One promising line is to assume not only that some laws exist, but that laws of a special form exist. We will consider six possible forms of laws and hence six new kinds of possible assumptions: (A) Causal laws; (B) time independent laws; (C) conservation laws; (D) minimum principles; (E) continuous laws; (F) laws involving a given number of dimensions. Of course, for any one of these six forms we could introduce an assumption of any of the three above types. Hence we could get 18 different types of assumptions. For the sake of simplicity we will here consider only assumptions of the first type, that is, assumptions about the possibility of learning laws exactly. We will show that of the six forms mentioned above, four lead to empty assumptions while the last two lead to genuine factual assumptions.

It will help in these considerations to illustrate our points with a specific example. Let us consider the law governing a stone that is thrown into the air. If we throw it from a height of h feet with a velocity of v feet per second, then its height at future moments is given by the formula $d = h + vt - \frac{1}{2}gt^2$. In this t is time measured in seconds, and g is a fixed number (approximately 32.2). This law, essentially due to Galileo, is of considerable interest since, if we neglect air resistance, the law applies to any physical object, be it a heavy stone or a feather. We will eventually show that this law can be rewritten in any of the four forms A through D.

The search for causal laws is deeply tied up with our subconscious tendency to recreate the universe in our own image. In our lives we search for motivating forces for all our actions. We say that a man is irrational unless he has had a "good reason" or a "good cause" for what he did. We are constantly stating what caused us to act in the way we did. "I went to the ball game because my father taught me to love the Dodgers." "I voted the Democratic ticket because of the speech I heard in October." "I believe in the U.N. because I believe it is our only hope for peace." In all three cases something happened in the past that caused the action in the present. What typifies causality is the fact that an event taking place at some time somehow brings about an event at a later time.

We demand that nature should follow our example. If anything at all takes place we immediately search for the cause in a previous chain of events. Why does the tree fall? Because we cut it down. Why does a balloon rise? Because air pressure lifted it. Why does wood burn? Because oxygen combined with it. In each case we search for an event which brought about the result. Subconsciously we feel that nature was forced by its past history to take a certain course of action, the same way as we as individuals are forced to take certain steps. When primitive people personify nature we laugh at them. Yet in many ways we are much more laughable, since we ought to know better.

One of the foremost British philosophers, David Hume, deserves credit for first pointing out that our attitude toward causality must be changed. Let us consider a typical example. We pull the trigger of a rifle and there is a loud noise. The cause is the pulling of the trigger and the resulting noise is the effect. If this happened but once in the history of the universe we would never associate the two events with each other. The reason we say that the former caused the latter is that we have often observed these two events together. Whenever we pull the trigger of a rifle (naturally a rifle in good working condition, loaded, and without a silencer), there follows a loud noise. But this sentence contains everything that is of factual significance. Whenever one event takes place it is always followed by an event of the second kind. If we add anything to it about one causing the other one, there is certainly nothing in our experience that warrants the assumption. The statement that an event of type A is always followed by an event of type B is a true general proposition and hence is a law of significance to Science. For example, if we pull the trigger in the future under the proper circumstances, we can predict that a loud noise will follow. There is no need whatsoever to assume that some mysterious force connects the two events. And even if there were such a force, if we cannot observe it in any way, we do not care about its existence. Hume does not object to calling one event a cause and the other one the effect, as long as it is clearly understood that this means no more and no less than that A is always followed by B.

What typifies causal laws is that, if we are furnished certain information about the present moment, the law will give us information about the future. This, of course, is of tremendous significance for Science. There is a second point of view from which causal laws are very important. They are the key of the division of labor between the theoretical scientist and the experimentalist. It is the role of the theoretical scientist to find causal laws, whereas it is the role of the experimenter to furnish data about the present. Combining the labor of these two types of scientists we are furnished information, not only about the present but about the future as well.

Let us now return to our example. The law we stated governing the tossed-up stone is in explicit form giving us information about past, present, and future. But it is easy to change this into a causal law. Suppose we state merely that the acceleration is equal to $-g$.

(The acceleration is the rate at which the velocity changes. The minus sign indicates that as the stone is flying upward its velocity is decreasing, while as it falls to the ground it falls faster and faster.)

It will be objected that this is not a causal law, since no "cause" has been found. But nothing is easier. We will say that the acceleration is caused by a "gravitational force," that is, by a force that causes the gravitational acceleration. How do we observe the force? By observing the acceleration. How strong is the force? Just strong enough to bring about the acceleration. This may seem artificial, but this is exactly what Newton did. In every type of motion we can observe the acceleration, *a,* and the mass of the moving particle, *m.* He then postulated a force causing the acceleration, of strength *ma.* Since in many types of motion the acceleration is inversely proportional to the mass of the moving particle, this gave Newton a constant force governing the motion. In our example even this satisfaction is lacking. The acceleration is independent of the mass of the falling body! But there is no difficulty in overcoming this; if the acceleration is constant, then we need simply postulate that the gravitational force is proportional to the mass of the falling body. Together with the fact that the force has strength *ma,* this says no more and no less than that the acceleration is a constant. The only possible justification for this bit of sleight of hand is that we demand that our laws be in the form of causal laws.

From this new form of the law alone we can say nothing at all about the position of the stone at any given moment. If we carry out a simple mathematical operation, namely integrating twice, we arrive at the explicit form of the law we started with, but with two important gaps. We do not know the value of *h* and we do not know the value of *v.* These values must be found by observation. Once these two numbers are found we have our original version reinstated, and hence we can have complete information about the flight of the stone. No matter how many stones we wish to study, for any stone we will need only two numbers. If we want to study a million objects thrown up into the air, the observer will have to get us two million numbers. As far as laws are concerned the same causal law will suffice. We have said that a causal law determines the future given the present. The two million numbers are our description of the present. The causal law does all the rest of the work for us. We might picture the process intuitively as follows: The given numbers tell us where the stones are at this moment and how fast they are moving. The causal law tells us then what will happen a brief time interval later. And given the information about this moment the causal law again tells us what happens after a short time has elapsed, etc. Putting all this information together we get a complete picture of the future.

We now see the great advantage of using this particular form for a law. Instead of having separate laws for stones thrown from different heights and with different velocities, we need but the single, simple law that the acceleration is equal to $-g$. All the rest is left up to the experimentalist. The more data he collects about the present, the more we will be able to say about the future. But it is also clear that this is but a convenient form for the law; it is not in any sense a special kind of a law. We had no difficulty at all in changing our law into a causal law, and it can be shown that there is no difficulty in this in any conceivable law. Therefore the so-called assumption about the existence of causal laws is not really an assumption but a convention as to the form to be used for laws of nature. We will later see that assumptions of the form B, C, D also are no more than conventions.

Naturally, conventions are very useful in any undertaking. The danger is in mistaking a convention for a factual assumption. There have been many philosophers who have tried to take the fact that scientists use causal laws and draw far-reaching conclusions from them. It is clear that the only legitimate conclusion to be drawn from the existence of causal laws is that scientists like to put their laws into causal form.

It has been maintained that the principal competitor of a causal law is a teleological law. The latter is a law which determines the present, given the future—rather than determining the future. A great deal of time and space in publications has been devoted to a discussion as to which type of law is superior. I do not want to leave this topic without considering the question.

Let us construct a causal law. We will choose the weather as our subject, and simply consider whether it is "good" or "bad." As the "past" let us consider the last three days. The law states that if of the last three days *exactly two* were good, then the weather remains the same as yesterday's; otherwise we have a change in the weather. We will see that this is a complete causal law as far as the goodness of the weather is concerned.

Suppose that we have had two good days followed by a bad day. We can symbolize that by "GGB."

Last three days.	GGB
Exactly two days were good, so the weather stays the same (i.e., bad).	GGBB
Now we see that of the last three days only one was good—a change is due.	GGBBG
Same as above—another change.	GGBBGB
Same as above—another change.	GGBBGBG
Now two of the last three days have been good—no change.	GGBBGBGG
Same as above—again no change.	GGBBGBGGG
Now all three days have been good, hence we are due for a change.	GGBBGBGGGB

We have now returned to our original case, and from here the weather repeats itself. We are led to a seven-day cycle. The information needed, namely the weather of the last three days, is the initial data, and everything about the future follows from this by means of our miniature law.

Can we say anything about the past? We again start with GGB. Let "X" stand for the weather of the day before, giving the sequence XGGB. Since XGG was followed by a change in weather, XGG cannot have *exactly* two good days in it. Hence X = G. Let "Y" stand for the day before, in the sequence YXGGB, or YGGGB. YGG is followed by no change; hence exactly two of these three days must have been good. Therefore, Y = B. And so on. Given the initial data, our law determines not only the future, but also the past. This is true also of the differential equations that express the causal laws of physics. What does this do to the old philosophical controversy about whether laws should determine the future on the basis of the present, or determine the present on the basis of the far future? The entire controversy seems to have lost its meaning.

OTHER FORMS FOR LAWS

Let us next consider the proposed assumption that laws should be independent of time. The difficulty with this assumption is that it is not at all clear in what sense a law must be independent of time. If we ask the scientist just what he means by this, a typical answer might be to point to certain features of a law which are not influenced by time. In the gravitational law he might point to the constant g and indicate the fact that this is a universal constant and that it is independent of time. But suppose we then ask him what would happen if g *did* change with time? Suppose we found out that this is only approximately a constant, that is, that it is a number that changes very slowly in time. Would he then abandon the attempt to formulate such a law? The answer is certainly "No." What he would do is to search for the law governing the rate of change of g, incorporate this into the old law, and thus arrive at a new law for falling bodies. This is just what was done when we learned that g depended on the altitude above sea level.

A second suggestion is that somehow the law should not contain t, that is, time, explicitly. We may talk about changes with respect to time and various rates, but t should not enter the equation, expressing our law directly. But there is no difficulty in changing our example to conform with this requirement. All we have to do is write the law in the form: The acceleration is $= -g$. This, as we have indicated before, is practically equivalent to our original law. It requires only that two numbers be found by experiments. And indeed this is a general feature of laws. Given any law whatsoever containing t explicitly, we can get rid of this by differentiating the law a sufficient number of times and combining the various derivatives to form a new law. This is the process by which an ordinary equation is replaced by a differential equation, a very familiar process in intermediate mathematics. The effect is always to replace an explicit formula by one that contains only rates of change. We must thus reach the conclusion that the proposed assumption that laws be time independent is empty of factual content. It is no more than a convention as to the form in which laws ought to be written.

Another proposed assumption that occurs very frequently in the literature is the assumption that Science is based on minimum principles. According to this, many fundamental laws are of the form that some quantity is as small as possible, that is, a minimum. Again we find that there is no difficulty at all in changing our example into this form. Let us consider the rate of change of the acceleration. Since in our example the acceleration is a constant, the rate of change of the acceleration is zero, which of course is its minimum possible value. (We are ignoring the question of whether the acceleration is positive or negative.) Thus we see that our original law can be replaced by the assumption that the rate of change of the acceleration is a minimum. There is a more drastic change possible by which our law can be changed to a minimum principle. We know that when there is no force acting, our stone will move in a straight line. Instead of introducing a force that pulls or pushes it out of this straight-line motion, we could change the entire geometry of space in such a way that the path the particle takes is the straightest possible path, known as a geodesic. This very far-fetched change underlies the General Theory of Relativity and has proved to be tremendously fruitful for modern physics.

Let us now consider a general type of law. We again find that there is no difficulty in changing it to a minimum principle. Suppose that a law of nature describes how a certain process changes with time. Let us say the law is given by a function $F(t)$ describing how the quantity depends on time. We need one idea from advanced mathematics: we need the idea of a distance between two such functions. This can be expressed by the formula $\int_{-\infty}^{\infty} |F_1(t) - F_2(t)|\, dt$. We can then express the law by saying that its distance from F is a minimum. We again arrive at the conclusion that since any law can be brought into the form of a minimum principle, this is not really a factual assumption about nature but again a convenient convention as to the form in which laws are to be written.

The assumption that occurs most frequently in the modern physics text is that nature obeys certain conservation laws. Let us again consider our sample law. We could describe it by saying that the acceleration is conserved. Again if we turn to the most general kind of law we find that there is no difficulty in writing it in the form of a conservation law. Suppose that it is in the form of an equation. We simply write the difference of the left-hand side and the right-hand side of the equation and, since this is always zero, we can replace our law by saying that this new quantity is conserved. This may appear to be a very artificial device, but a brief look at the history of conservation laws will indicate that they hold not so much because of any attribute of nature, but because of a human desire for conservation laws.

Presumably the original concept of energy was that of kinetic energy, or $1/2mv^2$. The original law of conservation of energy simply stated that in a physical process the total kinetic energy is conserved. However, it did not take very long to find out that this is not the case in general. For example, if a stone is thrown up in the air it is not true that its kinetic energy is conserved. Since it may start out at a high velocity, then come to a standstill momentarily at the peak of its path, and then pick up speed coming down, the kinetic energy will first decrease, become zero, and then increase as it comes down. From our sample law it can be deduced that $1/2mv^2$ is equal to a constant minus mgh. We could at this point have said that energy is not conserved. Instead, what the physicist did was to bring the mgh to the left-hand side of the equation and say that $1/2mv^2 + mgh$ equals a constant, and hence that the left-hand side is conserved. He went further and changed his mind about what to call energy, letting the whole new left-hand side equal the energy of the particle. He then said energy is conserved. However, this is still not the end of the story. A number of other factors came in, each one showing that his energy as he had it defined at the moment is not conserved. In each case rather than abandoning the law of conservation of energy, he changed the concept of energy.

Around the turn of the century perhaps the only quantity that was definitely distinct from energy was that of mass. On one hand, there were solid substances in physics which had mass; on the other hand, there were various types of energy. This final boundary too was broken down when it was found that one can change mass into energy. The law of change is of course the famous formula according to which a mass m can be changed into energy equal to mc^2, where c is the velocity of light. Even here the physicist did not choose to abandon the law of conservation of energy. Rather than that, he defined a new concept of energy which was the old one plus mc^2, usually called mass-energy, and then stated the law of conservation of mass-energy. We must again reach the conclusion that conservation laws exist not because of a peculiarity of nature, but because of a peculiarity of human beings, namely, that they like conservation laws.

This last example brings out clearly the difference between the form of a law and its factual content. Through the history of these laws of energy the factual content changed from time to time. However, the form remained the same. What makes the picture even more confusing is that not only was the form kept the same, but the name of the law continued to be "conservation of energy." Therefore, many people mistakenly thought that the law had remained the same. What actually happens is the following. The same equation is kept throughout the history of the conservation of energy law but a different interpretation is assigned to energy from time to time. This is perhaps the only clear-cut example where the factual content of a law is changed not by changing the equation expressing the law, but by changing the interpretation of the concepts entering into the law.

I do not deny that it is possible to construct new types of assumptions which have factual content. It is also clear that any of the above assumptions could be strengthened to give them factual content. But the usual discussions take these assumptions in their present form and then try to draw far-reaching conclusions from them. This is dangerous. There is nothing remarkable about a girl having blond hair if she dyed it herself.

FACTUAL ASSUMPTIONS

We will now consider two examples of assumptions about forms of laws which are not empty of factual content. The first one will be the assumption that The Law of Nature or some laws of nature are continuous. This means that if we draw a diagram of a law, this diagram is represented by an unbroken line. We may, for example, take the temperature in New York City and imagine that a temperature chart is kept over a long period of time. We must of course imagine that the temperature is noted for every moment. The assumption of continuity would mean that this chart consists of an unbroken line.

This is a very interesting assumption. Although it has factual content, it is impossible to disprove the assumption. Since human beings can make only a finite number of observations, no matter how much time is available to them, we find ourselves in a position where we can always keep the assumption that laws of nature are continuous. This is easiest to see by thinking of the laws as represented by diagrams. The finite number of observations would be represented by putting down a finite number of points on a graph paper. It is then possible to connect these points by pieces of lines in such a way that the total graph will have no breaks in it. Since this can always be done for any finite number of points, we can never get conclusive evidence that a law of nature is discontinuous.

Why do we want to have continuous laws of nature? The answer is that we would have a much better chance of finding a continuous law than one that is not continuous. First of all, we know that the collection of all continuous laws is of the second order of infinity. Although this is still higher than the collection of those laws available to human beings, it is lower than the collection of all possible laws of nature. There is also a second reason, namely, that continuous laws are usually in some intuitive sense simpler than discontinuous laws. Therefore we would have a better chance of finding them. But here is also the danger in our assumption. Although it is true that we can always keep our assumption of continuity, the price we have to pay for this may be very high indeed. For example, suppose that up to 2000 A.D. a certain number observed in nature always turned out to be three. From 2000 A.D. on, this number was always five. We could represent this by a continuous curve on graph paper, but it would seem much simpler to represent it by two separate lines, a line giving the constant value three up to 2000 A.D. and a constant value five from then on. In other words, the danger is that while we can always insist on a continuous law, and while this law is *usually* simple, the continuous law *may* be more complicated than the discontinuous law suggested by the evidence. Therefore, while we can never disprove our assumption, we can make it implausible enough to reject it. Indeed, the evidence now available in atomic physics is such that many physicists believe that certain atomic phenomena are discontinuous.

A second example of a nonempty assumption is the assumption that the space we live in has three dimensions. (We are only going to consider spatial dimensions and therefore time will not enter this discussion.) Physical evidence convinces us that we live in a space of at least three dimensions. The emphasis in this is on the words *at least*. We cannot ever get conclusive evidence that there are no more dimensions. We can only say that everything we have observed so far can be explained perfectly satisfactorily with laws assuming only three spatial dimensions. Although it is very unlikely that we are ever going to get beyond this stage, it is not inconceivable that certain evidence would lead us to reject the assumption.

For this we must investigate some properties of four-dimensional space. The easiest way to do this is by analogy. We have to try to picture a two-dimensional world and then imagine our relation to such a simple universe. One phenomenon alone will illustrate my point. Let us suppose that a three-dimensional being is captured in a world of two dimensions. We might suppose that the surface of a large, open field was this two-dimensional world. The three-dimensional being is to be jailed in this miniature universe. The only kind of jail that can be built around him would be a large square with the prisoner in the middle. But since the universe here consists of only two dimensions, the square would have no height at all. Therefore our three-dimensional being would have absolutely no difficulty in stepping over the "wall" and escaping outside his jail. How would this appear to a two-dimensional being? He would see the three-dimensional person inside his jail. The three-dimensional being would disappear momentarily from his view, and then reappear outside his jail. It can be shown that a four-dimensional being would have no more difficulty in escaping from one of our mere three-dimensional jails. If it were a common occurrence that certain beings would be able to disappear from our sight and reappear outside closed rooms, it is very likely that we would reject the assumption that we live in a space of three dimensions.

These two assumptions are examples where we assume that laws of a certain form exist and the assumptions are not empty. The assumptions can never be proved completely. In one case, it was impossible and, in the other, highly unlikely that they would ever be disproved. But they do share the common property that, although the evidence is unlikely to be conclusive, it can be sufficiently heavy so that we would come to a rejection of one or both of these assumptions.

UNIFORMITY OF NATURE

Whatever else a scientist will tell us about scientific assumptions it is likely that he will assign first place to the assumption that nature is uniform. This assumption, when we analyze it, will suffer from the usual difficulty. The form in which it is found in the literature indicates that this is somehow to be an admirable trait of nature. Actually what the authors have in the back of their minds is an assumption which connects human limitations with nature. The assumption is not so much an assumption about nature as about our ability to comprehend it.

In what sense is nature to be uniform? We are told that nature tomorrow will somehow behave just like today. We are told that the assumption is needed to assure us that our past experience is a reliable guide to the future. In other words, somehow we are to put bounds on nature so that it is to obey the same laws tomorrow as it obeys today. Fortunately, this terminology has already been shown to be untenable. A law is no more than a description of what actually happens. It is difficult to see what we would mean by saying that a description of tomorrow is the same as the description of yesterday. Literally this is false. In the sense of being part of the same description there is no earthly reason why it should not be true.

Let us construct a simple example of this. Suppose we draw two graphs on paper. Then we are told that one of these two graphs represents nature up to today and the other one represents it from today on. There is no reason at all why the two graphs cannot be com' ined into a single one (see the illustrations). In one example the resulting description will be a continuous one. In the other case it will be discontinuous. But this is a question of continuity and has nothing to do with uniformity of nature. It is almost certainly the case that here again we are suffering from a subconscious feeling that nature is like a human being. It is a common human experience that a law that was obeyed by certain human beings or a whole civilization for a long time is altered in some manner. We somehow want to prevent nature from doing this. It is clear that this type of thinking will lead to nothing but confusion.

Although the question of whether nature obeys one or more laws is pointless, we do feel a certain anxiety about The Law of Nature changing rapidly and regularly in time. We are worried that under such circumstances it would be hopeless for us to find this law. An even simpler example can be constructed. In the illustration we see a law of nature consisting of three very simple pieces. If our entire human existence is restricted to one of these three pieces, it is impossible for us by any reasonable means to guess the existence of the other two pieces. In this lies, I believe, the key to this question.

We may think of The Law of Nature as describing a highly complex pattern. What we can see of it is a small restricted piece. The pattern extends widely in space and in time. We can see only a small region of space at any one time and our entire existence may be restricted to a small time interval. Seeing this small piece of the pattern we must guess what the entire complex pattern looks like. We must now correct the previous impression. It is not

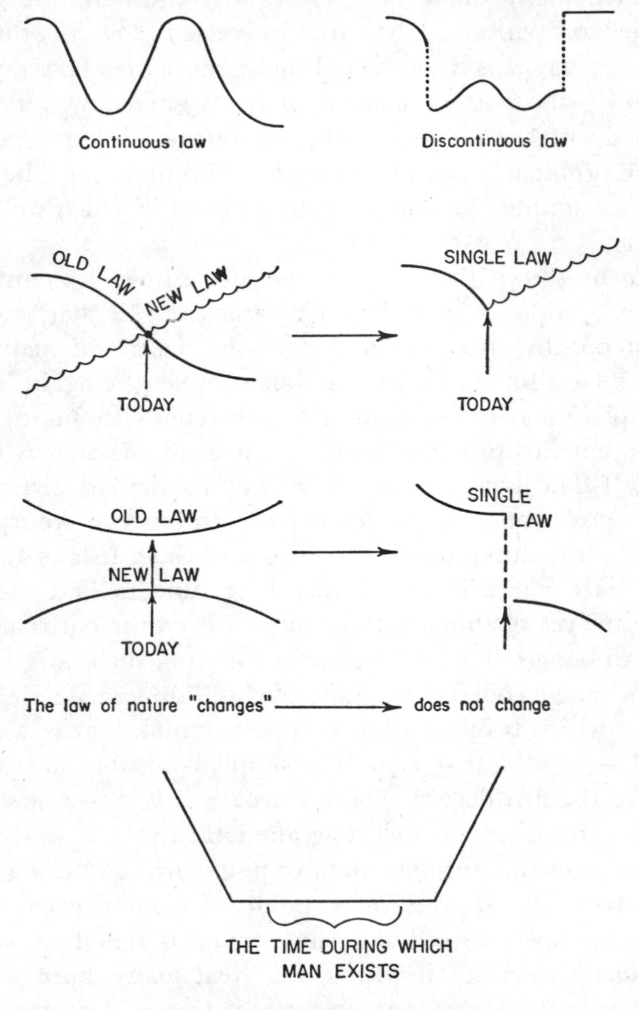

A law that we cannot find

so much a question of how rapidly the pattern is changing, but rather as to whether the changes we observe in our small piece of it are representative of the whole. We could easily construct an example of a rapidly changing pattern where we would have no trouble guessing the entire outlook. For example, suppose that it is a periodic pattern with a very short period. If the existence of human beings is long compared with this period we will see it repeated many times, and we will have no trouble guessing what the basic pattern of the entire universe is. On the other hand, if whatever happens in our small piece, no matter how rapidly or how slowly the pattern changes here, is somehow very simple compared with the whole pattern, then our task is hopeless. Therefore, the problem is one of the relative simplicity of The Law of Nature, its simplicity relative to the piece of it which we are able to observe.

It can be shown that, in scientific procedure, a scientist must consider hypotheses in order of complexity. He starts with the simplest possible assumption that fits his facts and maintains it until the facts force him to abandon it. Then he again looks for the simplest possible assumption consistent with his new facts. If *we* accept this procedure, the problem now facing us becomes clearer. If The Law of Nature or some particular laws are sufficiently simple compared with what we can observe, we have hope of finding it within a reasonable length of time. If laws are highly complex, then it will take a very long time to find them and, indeed, any reasonable length of time or even the entire existence of human beings may be too short for finding such laws.

Let us take a concrete example of this. Suppose we are looking for a law which is in the form of a polynomial, that is, a formula like $3x^3 - 2.3x^2 + x - 5.76$. The simplest possible such law will be one of the first degree, where x occurs only to the first power, and this can be represented diagrammatically by a straight line. Since we know that through any two points we can draw a straight line, it will take at least three points to convince us that our straight-line law cannot hold. Since actual observations are only approximately correct, it will take a great many more points, in general, to convince us that no straight line will fit the facts. If our law is of the one-thousandth degree we will need more than a thousand observations to arrive at it; and if we take into account again the inaccuracies of observations, it will take considerably more than a thousand observations. Thus we see that if a law, even as simple as a polynomial law, is of a very high degree, it may take a tremendous number of observations to find it.

In my belief, the assumption about "the uniformity of nature" really needed by the scientist is that our experience is representative of the behavior of nature in the large, or, to put it in slightly different form, that it is possible for us to learn about nature in a reasonably short time. An example of such an assumption might be: "It is possible for human beings to learn in ten-thousand years what The Law of Nature is." Of course there are many possible variants of such an assumption. Instead of The Law of Nature we may wish to learn only certain partial laws. Instead of ten-thousand years we may wish to specify shorter or longer periods or even indefinite periods. We could even substitute the entire history of human beings on the earth for such a time period. Again, instead of learning laws exactly, we could require only that we should learn them approximately or perhaps assume that we can learn them to better and better approximations with the passage of time. But whatever form of such an assumption we adopt, it will not be solely about nature, but will connect the complexity of nature with human ability to comprehend. It will have to be of the form that human limitations and the complexity of nature are such that it is possible for human beings to learn about nature.

Perhaps the scientist who most clearly understood the necessity for an assumption about the simplicity of laws was Albert Einstein. In an informal conversation he once told me about his thoughts in arriving at The General Theory of Relativity. He said that after years of research, he arrived at a particular equation which, on the one hand, explained all known facts and, on the other hand, was considerably simpler than any other equation that explained all these facts. When he reached this point he said to himself that God would not have passed up the opportunity to make nature this simple. Therefore he published this theory with an absolute conviction that it is correct. The dramatic success of The General Theory of Relativity bore out Einstein's faith.

Having spent some time in a discussion of the assumptions underlying Science, we must now take up the last prerequisite for a discussion of the scientific method. Even in the foregoing discussions it was very clear that the scientist does not deal with absolute certainties, but only with probabilities. This will be our next topic of discussion.

SUGGESTED READING

Complete references will be found in the Bibliography at the end of the book.

Causal laws.
- Campbell, Chapter III.
- *Margenau, pp. 389-426.
- Planck, Section 6.

Uniformity of nature.
- Mill, Book III, Chapter III.
- Cohen and Nagel, Chapter XIII, Section 9.

Number of dimensions.
- Abbott.
- Gamow [1], Chapter IV.

Orders of infinity.
- Kemeny, Mirkil, Snell and Thompson, Chapter II, Sections 12-13.

4

Probability

"It's really dreadful," she muttered to herself, "the way all the creatures argue. It's enough to drive one crazy."

OUR REAL AIM in this chapter is to discover how probabilities are used in science. In order to understand this important point we must have some understanding of what probabilities are.

THERE ARE TWO AND A HALF SCHOOLS OF THOUGHT

This unfortunately is not easy. It is not easy because even the "experts" are far from agreed as to the nature of probabilities.

Let us examine a few cases where probability statements are used in everyday life. Suppose we hear people make some of the following statements: "There is an approximately even chance that a child to be born will be a boy." "The odds against this atom splitting in the next minute are a billion to one." "The odds against rolling a natural are seven to two, though they vary for loaded dice." "It will probably be nice weather tomorrow." "He's very likely right on this." "On the basis of what I know I would bet two to one that he will marry the brunette."

You will probably have spotted that there is some difference between the first three statements and the last three. Although this difference cannot be made precise easily, it is about this intuitively felt difference that we find considerable dispute among the experts. If we are challenged to explain what we meant by the first three statements, all of us could come up with a fairly clear explanation. We know that after long and careful observations we found that approximately half the children born are boys and half girls. This is what we mean when we say that there is an approximately even chance that a child born will be a boy.

Similarly we could base our statement about atoms on a frequency count of splitting atoms, and we could base our statement about dice on noting the number of different ways that dice can come up and noting that the various ways come up equally often in the long run. In all of these cases we can make some statements about long runs and how frequently certain things happen. This is by no means easy in the last three cases. The last three examples deal not so much with what actually takes place in the world—that is, how frequently a certain thing happens—but rather with how sure we are that something will happen.

The first type of statement is beyond dispute. This kind of statement based on frequencies has been analyzed completely and an important branch of mathematics has developed treating these types of statements. This branch of mathematics is, of course, the Theory of Probability. There is general agreement that this theory has a firm foundation, that it is an extremely interesting and highly developed branch of modern mathematics, and that it is of tremendous importance in applications. So far all experts will agree.

When we come to the second type of statement, statements about how certain we are that something will happen, we find that there are two basically different schools of thought. The first school maintains that these too can be analyzed in terms of frequencies, whereas the second school believes that this is an essentially different concept. The reason this section is headed "There are two and a half schools of thought" is that the second school is split down the middle. Although members of the second school agree that there are two basically different concepts of probability, they disagree as to the nature of this new kind of probability. Many members of this school believe that the probability dealing with how certain we are can also be made as precise as the former concept, while other members of the school are firmly convinced that this concept is too vague ever to be made precise. One can find distinguished philosophers who belong to these various schools and branches of schools. The late Hans Reichenbach was a prominent advocate of the frequency theory. He firmly maintained that all probability statements can be analyzed in terms of frequencies. In the second school we find distinguished names like Bertrand Russell, Ernst Nagel, and Rudolph Carnap. Russell and Nagel have at various times emphasized the importance of probability statements about degrees of belief, but they seem to be quite firmly convinced that these are not open to precise mathematical analysis. Carnap, on the other hand, not only believes that such analysis is possible, but he has done a great deal to lay down the foundations of a new branch of mathematics which would deal with such statements.

In discussing Science, probability statements play three types of roles. First of all, statistical theories occur in many different branches of Science. Such theories make heavy use of probability statements. Secondly, all measurements are subject to error. The theory of errors is a branch of probability theory. Hence, no branch of Science can be entirely free of probability theory. Thirdly, when we assert a statement, we have to assign some degree of credibility to it, whether we explicitly state it or not. We must at least mentally note how certain we are of this assertion. Here probability theory enters for the third time. The last of these three uses clearly demands the second kind of probability statement. This is especially important when a scientist asserts a theory. After all, as long as he is only talking about the results of observations, he can be practically certain that his statements are correct. It is in theories that various degrees of doubt enter. We must also note that these statements are *about* scientific theories. They do not enter scientific theories or scientific predictions themselves. The first two applications of probabilities differ from the third on both of these points. They require probability statements of the first kind, analyzable in terms of frequencies; and they are all statements *within* Science rather than statements about Science.

For these reasons it is simpler to consider these first and we will be in a position to do that as soon as we have taken a more careful look at the nature of the frequency concept of probability. The third type of application will be considered in Chapter 6 where we discuss theory formation in Science.

THE FREQUENCY CONCEPT

Let us consider the simplest kind of probability statement involving frequencies. Suppose I take a coin out of my pocket and toss it up in the air. What is the probability that it will come up heads? Of course the probability is one half. Everybody knows that. But how many people really understand what this statement means? If we challenge the average person, he is likely to tell us that by this statement he means that if coins are tossed frequently enough and if they are not "loaded" coins, in the long run we will get approximately as many heads as we get tails. This is very interesting, but is it of any practical use? In order to apply this we must have a long run, we must be assured that it is a fair coin, and even then we will know only approximately what happens. Suppose I now toss a coin, how much information will it give me about what is likely to happen in the next ten minutes?

Let us ask that question more specifically. Suppose I grant you that this is a fair coin; that is, I am willing to grant you that it comes up heads with probability one half. What is the probability of getting two heads in a row? As long as you interpret the probability statement in the manner outlined in the previous paragraph this is an unanswerable question. You know that the probability of the first head in itself is one half, but we have no idea at all whether it is more likely that this would be followed by heads or by tails. For this reason we cannot tell what the probability of two heads in a row will be. Yet everyone knows that the probability of two heads in a row is one quarter. Something must have been lost in the process of clarification.

Let us ask again what we mean by a coin coming up heads with probability one half? We have already indicated the "important" part that in the long run approximately half the time it will come up heads. Have we left anything out? Can we say nothing at all about the way it comes up? Can we say nothing about whether heads are usually followed by tails or whether they are less likely to be followed by tails, whether they come in long sequences or runs of heads or can we make any statement about the order of events at all? The answer is that we cannot. This very fact is tremendously important.

The apparently negative statement that we can say nothing at all about the order in which heads and tails come up is expressed by saying that they come up "at random." If we analyze carefully the logical status of the two characteristics of probability statements we find that the fact that heads come up half the time is a very weak statement. The fact that heads come up at random is a very strong one. This will sound less paradoxical if we illustrate it in a concrete example.

We will prove that if you add randomness to the very weak statement that heads come up half the time, it will then follow that the probability of two heads in a row is one quarter, though this certainly does not follow without the extra assumption. But in order to do this we must put our assumption of randomness into a more useful form.

So far we have only stated negatively that we can say nothing at all about the order of events. But we could turn this around and state positively that any prediction we make as to the order in which heads and tails will come up will be false. This reminds us of an experience many of us may have had in trying to predict the behavior of a roulette wheel. As many people have found to their sorrow at Monte Carlo, whenever they believed they had an infallible system of predicting the behavior of this unpredictable wheel, they ended up losing a great deal of money. Monte Carlo has remained a profitable business enterprise because there is no way of predicting the sequence of numbers on a roulette wheel. This fact, that there is no infallible winning system at roulette, or at the game of heads or tails, is precisely what is meant by saying that the events occur at random.

Consider the various possible alternatives in tossing a coin twice. It could come up HH (twice heads) or HT or TH or TT. It is my claim that all these possibilities are equally likely, i.e., that each one has probability one quarter. Let us suppose that this is not the case. Say HT is more likely than HH. This would mean that whenever we have observed an H—i.e., heads coming up—we would know that on the next toss it is more likely to be tails than heads. But this would give us an infallible winning system. We would just wait until heads came up and the next time we would bet on tails. Doing this long enough we would clean up on the probability differential and become very rich indeed. Granted the assumption of randomness, however, there is no sure-fire winning system. Therefore, this proves that an HT sequence cannot be more probable than HH. Similarly we can prove that all the alternatives are equally probable.

Herein lies the secret of the strength of the assumption of randomness. There are many types of information which, if available to us, would allow us to win consistently. If there is no guaranteed winning system, we can conclude that these sources of information are not open to us, and therefore we can conclude in many cases that certain types of outcomes must be equally likely. For example, if we toss a coin ten times, there are 1024 different ways in which the sequence can come out. By an argument exactly like the previous one, one outcome is as likely as any other. Again, by applying the same type of argument, we can see that the outcome the tenth time is entirely independent of what happened before. If the first nine times the coin came up heads it is just as likely that it will come up heads as tails next time. We have often heard people in such circumstances say that "by the law of averages" it is pretty sure to come up tails the next time. But the law of averages is not a conscious, man-like agency that keeps a careful record of what has happened. We are again guilty of trying to rebuild nature in our own image. We must conclude that the phrase "by the law of averages" is almost invariably misused. If the law of averages really required that nine heads should generally be followed by tails, you could be quite certain that people would be sitting around in Monte Carlo waiting for long runs of reds and then winning large sums of money betting on black. Again, thanks to randomness, Monte Carlo is still a profitable business.

So far my illustration has shown only how, from a single assumption about the probability of heads, we can derive a large number of statements as to certain possibilities being equally likely. But suppose we want to make statements that certain events are more likely than others. This too can be done by lumping together various possibilities. Let us again consider the case where a coin was tossed ten times. Let us ask whether it was more probable that we get five heads and five tails or that we get heads ten times. The answer is that the former alternative is *much* more probable. Not only does this not contradict what we have said in the previous paragraph, but it actually follows from it. While it is true that one sequence of events is as probable as another one, we must remember that there is only one way of getting ten heads in a row, but there is a very large number of different ways of getting five heads and five tails. There are actually 252 ways of getting five heads and five tails. Since any one of these is as likely as getting ten heads, the probability of getting five heads and five tails is 252 times as probable as that of getting ten heads.

Even extremes of probability can be manufactured by lumping together possibilities. For example, if we toss a coin a thousand times and keep track of the number of heads and tails, we can predict with overwhelming probability, practically with certainty, that the outcome will be between four hundred and six hundred heads. The reason for this is that, if we write down all the possible ways in which the sequence of a thousand tosses can come out, we will find that most of them give us somewhere between four hundred and six hundred heads.

This concept of practical certainty plays a fundamental role in the scientific method. Since we have already hinted at the fact that no branch of Science can escape probability theory completely, we must also ask how we can ever test our predictions? If our predictions turn out to be but predictions of probabilities, then we can never state definitely whether an outcome is or is not what was predicted. Let us suppose that we predict that the coin will come up heads with probability 0.78, and it does come up heads. Does this agree with our prediction or disagree with it? Of course this question makes no sense at all. In this sense, if our prediction is in the form of a probability, we can never definitely disprove it. But here is where practical certainty comes to our aid. If we can predict that a certain event will take place with a probability so great that we are practically certain of it, then we will accept its failure to appear as definite evidence that somewhere we were wrong.

Probability theory delivers an additional very important mathematical result. We find not only that we can manufacture practical certainty out of equal probabilities, but that as the number of events we considered increases we can sharpen up our results as much as we wish. In the case of tossing a coin, for example, not only can we say that we are practically certain of getting between 40 and 60 per cent heads in a thousand tosses, but we can also say that if we toss a coin more than a thousand times we can make sharper and sharper predictions. As the number of tosses increases greatly, we can decrease the percentage margin we allow ourselves as much as we like. This will be illustrated in the next section.

STATISTICAL LAWS

Let us try to put ourselves into the shoes of a scientist who has carried out a large number of experiments and is now searching for a law which will describe the events which he has studied. He tries a large number of laws which suggest themselves and carefully checks each one against his experiments. But no matter how long and how carefully he searches he finds that none of the suggested laws agree with experience. At this point he is almost willing to give it up as a completely hopeless task. But just before he does he may have an inspiration and try to do exactly the opposite. Instead of trying to find a law that agrees with his experiments, he will try to assume that there is no such law. He will try to assume that the events occur completely at random, and if this assumption works he will arrive at a theory using probabilities—he will have constructed a so-called statistical law, which could more appropriately be called a probability law. In other words, statistical laws enter into Science when all the regular methods fail or when we are forced to confess complete ignorance.

In this case we try to make a virtue out of our shortcomings and change our ignorance into a postulate of "complete lawlessness" and then feel free to apply all of the powerful mathematical theory of probabilities.

If you are an observant and critical reader you will catch me in a contradiction. I here state that an event obeys no laws and yet I took great pains in Chapter 2 to point out that such a possibility does not exist, since it is self-contradictory. What I must mean is that there is no law statable in a human language which describes the events correctly. This is the essence of the now accepted definition of randomness, due to the American mathematical logician, Alonzo Church.

I will choose a timely example to illustrate the nature of statistical laws. The example I have in mind is the disintegration of ordinary uranium as found in nature. We know that uranium left to itself will slowly disintegrate. However, this process proceeds at such an extremely low rate that it takes about a billion years for an appreciable portion of the uranium to break up. (It is this fact that has been used to date some of the oldest rocks.) Given a pound of uranium, we know at least roughly how many atoms it contains. We know what portion of these atoms will disintegrate in a year, fairly precisely. But we have no idea at all which particular atom is going to split at any given moment. We have no idea whether a selected atom will split within a millionth of a second or will be there for billions of years. No doubt there is a precise law governing the order in which atoms of uranium in a given material break up. But this law is entirely unknown to us. In such a situation our best alternative is to assume "complete lawlessness," that is, that the atoms that split are selected at random.

We can now apply the lesson we have learned from studying the tossing of coins. If this is a random process and if we know the probability of an atom splitting in the next second, we should be able to predict approximately how many atoms split in any one second just as we predicted that we get approximately five hundred heads in a thousand tosses. This number turns out to be of the order of magnitude of five million atoms a second (we could make this number fairly precise). We also know, however, that in a random process we cannot predict exactly how many atoms will break up, just as in tossing a coin we could only be certain that we would get somewhere between four and six hundred heads. Here, however, we are dealing with much larger numbers. It is still true that we cannot tell anywhere nearly what the exact number will be. No matter what prediction we make, we may be off by a few thousand; but we can be practically certain that the predicted number will not be off by as much as 1 per cent and that should be good enough for any practical application.

The key law of probability involved here, called the Law of Large Numbers, is quite simple to understand. Let us say that we want to be 99 per cent sure of our prediction. That means that we must allow ourselves a margin of error. This margin will depend on the number of events we are dealing with. As the number of events increases, the size of the margin must also increase, but the relative size (percentage contained in the margin) decreases steadily! For a hundred tosses of the coin we need a margin of about 15, for a thousand a margin of 50, for ten thousand a margin of 150—but the percentage margin goes from 15 per cent to 5 per cent to 1½ per cent. (These are approximate figures.) In the case of the uranium we are dealing with a trillion trillion atoms. And although only a negligible number split in a given second, this negligible number is in the millions. While the margin of error must be some 7000 atoms, this is only a fraction of a per cent of the predicted total.

This example is quite typical of statistical laws. We make the assumption of randomness where our ordinary methods have failed. Then, given a sufficiently large number of events to which to apply our theory, we can make predictions which are practically certain to be correct. If we find that the predictions are not verified through experience, we can be practically certain that our theory was wrong. Although this state of affairs is not quite as satisfactory as a precise theory, where everything is black or white, we at least have very dark shades of gray and extremely light shades of gray—which is almost as good.

Actually, the laws discussed so far are based on the simplest type of probabilistic assumption. More modern branches of probability theory, like the theory of Markov chains, will allow much more sophisticated assumptions. For example, we may assume that a gas must be in one of a very large number of states, and that the probability of being in a given state is determined by the last state of the gas. This leads to a Markov chain, in which predictions can be made about the long-range behavior of the gas. For example, we can predict with high accuracy the average number of times that the gas will be in each state.

THE THEORY OF ERRORS

The second source of probability statements within Science is in errors of measurements. In the previous section we distinguished statistical laws—that is, laws incorporating probability statements —from precise ones. But it must not be thought that if we employ precise laws we are entirely free of probability statements. Indeed, as we have indicated above, no branch of Science is entirely free of these.

It may be advantageous to start out with a precise theory, but theories are of no use unless they are tied up with experience. To tie up a theory with experience we must have some initial data and we must draw some conclusions in the form of predictions. Although the theory may be precise, our contact with experience is always subject to error. Since there are contacts before the theory is applied, and contacts after the theory is applied, there are two sources for the entrance of probability statements.

Let us suppose that our contact is of the form of measuring lengths by means of a yardstick. The yardstick is marked to the nearest sixteenth of an inch. If we are careful in laying down the yardstick and we don't have to use it too often we may be quite sure that our measurement is correct to within one eighth of an inch. But we can also be quite sure of having committed an error of at least one one-hundredth of an inch. To be more precise, we can say that errors of the order of magnitude of a hundredth of an inch or two or three hundredths of an inch are highly probable. Larger errors get less probable until finally we can be practically certain that we will not commit an error as large as one eighth of an inch. This situation is represented by the familiar bell-shaped curve (see the illustration). While the numbers here quoted apply only in the specific case, the shape of this curve and the facts it represents are typical of what we run into in making measurements, and the reason we know that they are typical is because the theory of errors is a branch of probability theory and hence can be analyzed quite precisely mathematically. As such we find that the analysis of the previous section applies. If our prediction turns out to be exactly correct, we must attribute this to sheer luck, as coming out with exactly five hundred heads in a thousand tosses is purely a coincidence. We can be quite certain that there will be a small error, but we can be equally certain that the error will not be very large.

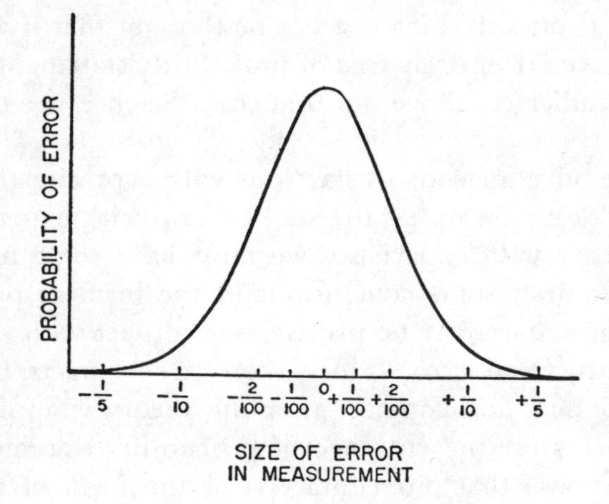

Suppose that we apply this to a practical problem. We are to fire a cannon at 40 degrees elevation. To simplify the problem, let us suppose that we know the initial velocity of the shell precisely, and we are to determine how far away it will land. With some oversimplification we could state that the initial data are given entirely by the degree of elevation of the cannon. Verification at the other end consists in measuring the distance from the firing place to the hit. Both of these measurements are subject to error. If we looked only at the measurements of length involved in finding out how far away the shell landed, we might be tempted to say that it is quite easy to measure this distance to a sixteenth of an inch. It may be somewhat painstaking labor, but if we are willing to do it, we could string out a rope to make sure that we have a straight-line measurement and very carefully apply a yardstick to it time and time again. Or we could use some other measuring device of even greater accuracy. However, this is only half our problem. After all, we cannot be certain that the cannon was elevated exactly 40 degrees and at this end we are in serious difficulties. A 1 degree difference in elevation may be a relatively small matter for something as bulky as a cannon, but we find that such a difference in elevation could easily cause an error of 10 feet in our prediction as to where the shell should fall. The way we arrive at this figure is by a combination of the initial data and the theory. If we know that the initial elevation is subject to an error of at most 1 degree, we can say that it is somewhere between 39 and 41 degrees. We can calculate from the theory for all such initial elevations just where the shell will fall, and then the best we can predict is that the actual outcome will be somewhere within this range.

Of course no actual experiment is that simple. We have had to assume that we knew the original velocity of the projectile as precisely as we wished. We also ignored winds and air resistance, both of which are important factors and could throw off our prediction by several yards. Perhaps the only statement we could assert with practical certainty is that the shell will fall within 100 yards of the point predicted. If we analyze the final prediction carefully, we will note that we again have a bell-shaped curve with its peak somewhere near the predicted point of impact, indicating that we can be pretty sure that we will hit near this point, though not exactly at this point, and that we can be practically certain that the actual outcome will not be off by as many as 100 yards.

We find ourselves in the strange situation that, while there is a great deal of theoretical difference between precise laws and statistical laws, as far as practical predictions are concerned, they are of exactly the same form: in either case the predictions must be of the form of probability statements. In either case we find that we can give some value to which the actual outcome will be near, but that we must allow small deviations from this, and we might in a few instances have to allow fairly large deviations as well.

In classical physics, that is, physics before 1900, there was some consolation in this picture. Although the physicist had to admit that his predictions were never precise, he at least consoled himself with the belief that he could come as near precision as he wished (this is like the closer and closer approximation of the square root of two we discussed in Chapter 3). However, in modern physics even this consolation has been taken from us. Due to the celebrated Heisenberg Principle we know that there is a limit to the precision with which we can accumulate our data. For example, if we measure the velocity of an electron at all precisely, say within 10 per cent, then our measurement of its position will be off very badly. The error in position, though infinitesimal, will be so large compared to the size of the electron that it would compare to an error in human measurements of the order of 1 mile. According to the Heisenberg Principle there is absolutely no way of breaking this gap, and the principle is very well supported by many experiments. Hence we are forced to conclude that no matter what type of theory we have available, there is an absolute limit to the precision with which we can make predictions.

We arrive at a very complex picture of checking theoretical predictions against experience. When a new scientific theory is published the impatient newspaper reader expects headlines next morning as to whether the theory has checked with experience. Or perhaps he is willing to allow some time for the preparation of an experiment, but as soon as the experiment is carried out he expects it to yield a definite yes or no answer. The situation is by no means that simple.

Our best analogy here is to compare the testing of a scientific theory to the checking of a roulette wheel. In a scientific theory we have predicted the result of a certain measurement and we have had to allow errors of certain magnitudes with certain probabilities, as shown by a bell-shaped curve. In the roulette wheel we predict that each number comes up with given frequency and that each comes up at random. To test the latter prediction one would have to accumulate a tremendous amount of data, and one would have to check carefully whether discrepancies are of the order of magnitude to be expected, that is, whether the outcome is fairly probable or not. Anyone who has visited a large casino like Monte Carlo will have found many people sitting around roulette wheels doing nothing but testing the scientific hypothesis that the roulette wheel is really a fair wheel. Many gamblers have regretted that they have jumped to a conclusion on the basis of but a few days experience.

What kind of experience would one accept as conclusive? Certainly the fact that long runs appear is not conclusive. By the theory of probability we know that we must expect runs of arbitrary length if we are patient enough. The question has to be how frequently these occur and how well they fit into the overall pattern of what has occurred. The methods of testing these have been developed in modern statistics. They are difficult to apply and impossible to summarize in a short space.

By analogy we can conclude that in testing a scientific hypothesis we must have great patience and accumulate a vast amount of observational data. The fact that a small amount of data does not definitely disprove a hypothesis should not allow us to jump to the conclusion that we have found a correct scientific theory. Neither should an apparently improbable outcome discourage us until we have reproduced this a number of times. A single run of red twenty times in a row does not prove that the wheel at Monte Carlo has developed a defect. But ten such runs in one evening would certainly induce the management to discard the wheel. Only if the outcome of experiments has repeatedly turned out to be consistent with our theory—that is, not too improbable according to our theory—will we feel any confidence in it. Only if *highly* unlikely events have occurred will we start looking for a flaw in our argument.

THE GREAT DISPUTE

As an application let us consider what is no doubt the hottest dispute in modern physics. Admittedly this is a dispute within Science, and not a dispute about Science, and hence does not properly belong in this book. However, it is so directly an application of the principles so far considered that I feel justified in saying a few words about it.

The newest and perhaps most celebrated branch of modern physics is Quantum Theory. This is a statistical theory. Some of our leading scientists have remarked on the fact that several times in the history of Science a new theory started out as statistical and was later replaced by a precise theory. From our analysis of the nature of statistical theories this is not too surprising. If in many cases a statistical theory arose because we were unable to find a precise law, it is quite natural that, should we find later on that such a law does exist, we replace our statistical theory by an exact law. The dispute concerns the question as to whether this will ever happen to Quantum Theory.

We know of course that a precise law must exist. The question therefore is whether a precise law capable of replacing Quantum Theory is statable in a human language. If the answer to this is "No," then our assumption of randomness is completely justified, and indeed Quantum Theory is the last possible word in microphysics. Hence the problem is one of the human ability to express complex laws. However, the disputes—and there were many and quite heated ones—certainly were not in these terms. We find that the vast majority of contemporary physicists, led by no less a man than Niels Bohr, maintain that the Heisenberg Principle somehow indicates that an exact theory cannot exist. As we have noted previously, this principle states that it is impossible to accumulate initial data and to verify any prediction more precisely than that given by specified bounds. From this, many physicists have arrived at the conclusion that a precise theory is impossible. It is not for the philosopher to try to guess whether the conclusion is right or not. It must however be pointed out that such a conclusion does not follow by pure logic from the Heisenberg Principle.

It is interesting to note that, although the opposing camp is very small, it includes a number of our greatest twentieth-century physicists, and some of the people who have done most for the development of Quantum Theory. In recent years in this group were men such as Einstein, Planck, and Schroedinger. These men have maintained in various forms that Quantum Theory, though certainly a great contribution to human knowledge, will be superseded eventually by a precise overall theory of microphysics. Of course, I must state right away that I am not trying to settle a scientific question by considering who is arguing on which side. Scientific matters are never decided by majority vote. But it is equally important to point out that scientific questions are not settled by a priori arguments, whether these arguments are presented by philosophers or scientists. This is a purely scientific issue—an issue which must await the development of Science.

SUGGESTED READING

Complete references will be found in the Bibliography at the end of the book.

Nature of probability statements.
Russell, Part Five.
Carnap [2], Chapter II.
Reichenbach [1].

Probability theory.
Kemeny, Snell and Thompson.
Nagel [2].
*Feller.

Theory of errors.
*Arley and Buck, Chapter 11.

Heisenberg Principle.
Eddington, Chapter X.

PART TWO

Science

5

The Method

"I should see the garden far better," said Alice to herself, "if I could get to the top of that hill: and here is a path that leads straight to it—at least no, it doesn't do that*—but I suppose it will at last. But how curiously it twists! It's more like a cork-screw than a path! Well* this *turn goes to the hill, I suppose—no it doesn't! This goes straight back to the house! Well then, I'll try it the other way."*

THE FIRST FOUR CHAPTERS covered the preliminaries: what we must know before we can understand Science. Now we are ready to turn to a study of our proper subject matter—the nature of Science. The most characteristic feature of Science is its method, and this is the first thing we want to study. I will maintain the thesis that there is one basic method common to all of Science, and I will try to show just what that method is.

THE CYCLE

As Einstein has repeatedly emphasized, Science must start with facts and end with facts, no matter what theoretical structures it builds in between. First of all the scientist is an observer. Next he tries to describe in complete generality what he saw, and what he expects to see in the future. Next he makes predictions on the basis of his theories, which he checks against facts again.

The most characteristic feature of the method is its cyclic nature. It starts with facts, ends in facts, and the facts ending one cycle are the beginning of the next cycle. A scientist holds his theories tentatively, always prepared to abandon them if the facts do not bear out the predictions. If a series of observations, designed to verify certain predictions, force us to abandon our theory, then we look for a new or improved theory. Thus these facts form the fourth stage for the old theory as well as the first stage of the new theory. Since we expect that Science consists of an endless chain of progress, we may expect this cyclic process to continue indefinitely.

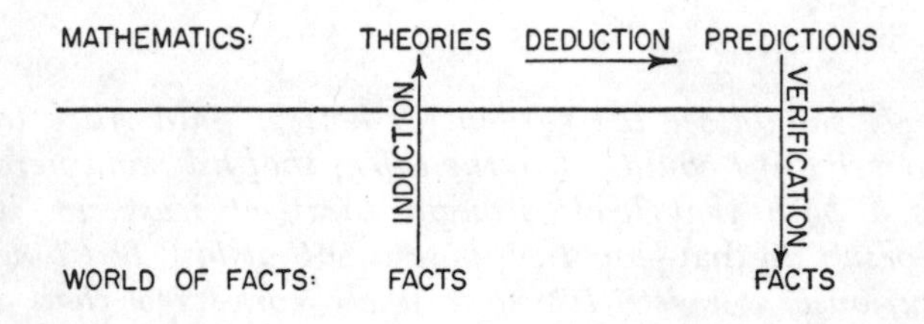

The horizontal line in the diagram separates the world of the experimentalist, the universe of facts, from that of the theoretician, the world of Mathematics. In the world below (the line) we find men peering through microscopes, while above we find an endless string of mathematical formulas. What will interest us most in this chapter is the way we proceed from stage to stage; accordingly we will study three steps. The first step carries us from the original observations to the theories. This is known as "induction," or the formation of theories on the basis of factual knowledge. As we have seen, this means that the scientist finds a mathematical formula which he can interpret to suit the facts that he is trying to incorporate in a theory. Then he asks himself the question: "Is this really what I want?" And he is forced to go back to the world of facts to check his construction. But you cannot check a general law directly; you must first ask what it tells you about particular facts. You cannot observe that the sun rises every day throughout eternity; what you can observe is that it rises today, and that it rises tomorrow, and the next day, etc. Any (finite) number of these can be checked. So the scientist must get from his general laws a prediction as to what will actually happen, say, tomorrow. This step is accomplished by "deduction,"

by logical analysis of what the general law says about a particular event tomorrow. Then he is ready to return to the facts, and see whether he was right in his prediction. This third and final step, consisting of experiments or observations, is the "verification" of the theory.

As an example of this cycle, I will cite one of the most dramatic chapters from the history of Science.

Our story starts in the year 1820. The French astronomer, Alexis Bouvard, was a little-known scientist whose contribution consisted in a painstaking charting of the paths of the planets. He was especially interested in the three large outer planets, Jupiter, Saturn, and Uranus. Bouvard was performing the very important task of accumulating more factual knowledge, enabling us to check and recheck the accepted theories. Newton's theory was accepted without question as a complete explanation of planetary motion. It was, therefore, a great shock that the observed positions of Uranus did not agree with the predictions. The deviation was small, but it was more than one minute of arc—which could not be put down as an error of observation.

This is the last step in one cycle of the scientific method. Many data were accumulated, primarily by Tycho Brahe. Kepler, Galileo, and Newton succeeded in formulating a good theory. Thousands of predictions were made on the basis of this theory, and until this time all of them were verified. But a single (carefully checked) incorrect prediction is sufficient to force us to modify our theories.

Yet all that we have really shown is that some theory, now accepted, is wrong. We still have a choice as to which theory to abandon. In this we must remember not only general theories, but particular ones; for example, our assumptions as to how many planets there are. Newton's theory was so well established that scientists would rather have abandoned any other part of the accepted body of knowledge. Hence, soon they came to guess that they must have been wrong in assuming that Uranus was the outermost planet.

A new, modified body of theories was formed by assuming that there was a planet beyond Uranus. But this was not enough to explain the observed facts. One had to show that a planet of the right size in the right place would account exactly for the observed deviations. Since the size of Uranus was known, and since Newton's theory as to the strength of attraction between planets was still accepted, it was a problem of pure mathematics to deduce the size and position of the hypothetical planet—or to show that a new planet cannot explain the observations.

The French mathematician Leverrier was the one who succeeded in solving this problem. With the mathematics known in those days this was a very difficult problem requiring considerable originality. Today it would be a routine assignment. Leverrier was able to determine both the size and the position of the unknown planet, which enabled astronomers to look for it.

We might be tempted to say that this was unnecessary. Why couldn't astronomers simply scan that portion of the sky until they found a new planet? The answer is that planets are not at all easy to locate. A planet, unless it is very near us or very large, looks no different from the billions of stars. It can be distinguished only by its path. The stars appear to revolve around the earth as if they were attached to a glass sphere (as the ancients thought they were), while planets move in a more irregular path. We would have to chart the position of all the stars in a region and follow these over a period of weeks until we spotted one whose position relative to the rest is changing. This would be a nearly hopeless task.

But with Leverrier's calculations in hand the Berlin Observatory knew the exact position of the sky and the magnitude of the "star" to look for. This simplified their work sufficiently so that in a brief period of intense observations they verified the existence of the hypothetical planet. The newly discovered planet was named Neptune.

This completed the most recent cycle of scientific method. The previous cycle was finished by Bouvard failing to verify predictions. The inductive step was formulated by several scientists who proposed that we modify our theory as to the number of planets. The deduction was Leverrier's; through a difficult mathematical argument he predicted the size and position of Neptune. The verification was accomplished at the Berlin Observatory. The cyclic nature of the process is further emphasized by the fact that it was repeated along similar lines in the twentieth century, when Pluto was discovered.

FACTS VS. THEORIES

Before we discuss the three steps of the scientific method, I must say something about the two "worlds" of the scientist. One is the everyday world we are all familiar with, only the scientist's familiarity with this world derives from careful, accurate observations. The other is the mysterious, fascinating world of the theoretician, the world of ideas, the world of mathematical formulas. Establishing a connection between these two worlds is one of the most difficult tasks a scientist must face.

We would like to think of a scientist as starting with "hard facts," and building theories on these. But I doubt that we can state a fact entirely divorced from theoretical interpretations. You might feel that when you see a table, you have a hard fact, but you have actually made use of certain theories you have so thoroughly accepted and assimilated that you use them subconsciously. I certainly do not deny that it is a "hard fact" that you have a sensation which we commonly describe as seeing a table, but this is not all that you mean when you state that you see a table. Suppose I ask you whether you could stick your fist through the object you see; you indignantly reply that the answer is most certainly "No"; after all you just said that it was a table that you saw. But there is nothing in your visual image that makes it logically certain that you see a solid object.

As a matter of fact, under certain circumstances, as in dreams and mirages, you can "put your hand" through a seen "table." It is a theory based on past experience that certain visual images are associated with solid objects. You will also assume that the top of the table looks four-sided from all points of view, but that while it looks like a rectangle from above, the angles will vary as you walk around it; in other words you assume certain primitive optical laws. While there are primitive "hard facts" in your experience, your report of your experience always contains an interpretation of what you think you saw. Sometimes the laws that you assume are far from elementary. When the biologist looks through a microscope and reports seeing a minute living creature, he makes use of advanced laws of optics (in connecting up what he sees through the microscope with what there is) and of laws of Biology (in inferring that the image is that of a living being).

In order to make his facts as reliable as possible, the scientist performs experiments. He arranges a physical situation with specified details, and then he reports nothing but what his instruments "tell" him. It has sometimes been said that all of Science is based on pointer-readings. This is somewhat exaggerated, because the readings tell us nothing unless we know what the meter reads, but it does bring out an important technique. The situation is further complicated by the fact that the scientist himself may serve as the instrument—for example, when he counts the number of albino guinea pigs, or when he "records" the reaction of his subject to a question.

Remembering that this is an oversimplification, let us accept that the scientist starts out with reports of what there really is. (We will return to this problem in the Chapter 7 on concepts.) But what the theoretician gets out of this is a statement like: $x = 3.25, y = 2.97, z = -4.00, t = 49.32$; or he may end up with points of a graph, or a "table" (say of "yes" and "no" against various questions). These are the so-called facts which the theoretician must unite into a theory. His x may be a reading on a ruler, his y the height of a column of mercury on a fixed scale, his z the elevation above sea level measured by a complicated surveying technique, and t the reading on a clock. These are the technical representations of observations, which record one or a series of events. His theories, on the other hand, will be generalized mathematical statements. A theory may be an equation like $xy - z^3 = t - 21.35$, or it may be a complete graph, or a rule of how tables will look in all cases.

It will be convenient to speak of the mathematical record of a fact (like $x = 3.25$, etc.) as a fact. If we allow this, then the only difference between a fact and a theory is that a fact is something that we already know, while a theory also states things not yet observed (and possibly never to be observed). There is a second difference which holds in most, but not all, cases: A fact reports a single event, while a theory reports an unlimited, perhaps infinite, number of events. The reason this does not always hold is that the latter may be false. A fact is always a single thing, like "there is a sun in the sky right now." Although a theory is generally a statement, like "the sun rises every 24 hours," the statement "the sun will rise at 8:00 A.M. tomorrow" also has the status of a theory. The reason for this double possibility is that, while a genuine theory is universal (rather than particular, like a fact), it has logical consequences, which are particular statements. For example, the particular statement "the sun will rise tomorrow" follows from the universal "the sun rises every day."

After making these subtle distinctions, I will ignore them. I will normally use the word "theory" only to apply to universal statements. Hence there will be a twofold difference between facts and theories: Facts are known and particular (refer to a single event), whereas theories are universal and hence can never be known to be entirely true. It is because of this universal nature of theories, because they apply (in principle) to an infinity of events, that Mathematics is almost always indispensable for their statement.

So the scientist makes some observations (perhaps as the result of a planned experiment), and records these in the mathematical language devised by the theoretician. The theoretician tries to formulate a general mathematical proposition, incorporating these facts. Then he develops this theory mathematically, deriving certain predictions of facts. These predictions are, of course, still mathematical propositions, and must be translated back into everyday language before they can be checked.

Let us take an example. We observe a plane that has taken off for a nonstop flight around the world. We note its position at various moments, and record these. What we see is that the plane is over San Francisco at a certain time, over St. Louis at a later time, over Pittsburgh still later, over Newark still later. For simplicity let us record only its longitude and the time (in hours counted from take-off). The table shows that it is traveling at a constant speed (roughly) of about 10 degrees longitude in an hour.

Longitude	*Hours elapsed*
122°	0
90°	3
80°	4
74°	4¾

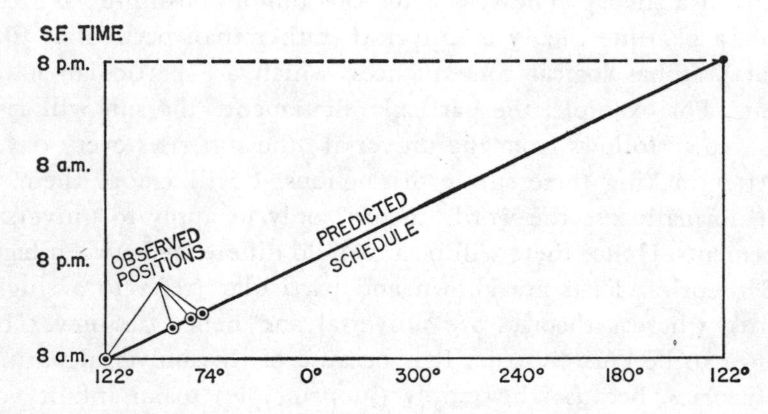

What we actually have done has been to plot the four reports as points of a graph, and then drawn the simplest smooth curve fitting these points well. It happened to be a straight line. This line is our theory; it is universal since it tells us where the plane will be at every moment from take-off to landing. From it we can read off that the plane will reach the original longitude in 36 hours, and hence we predict that (assuming that the plane left at 8:00 A.M. Monday) it will arrive back in San Francisco at about 8:00 P.M. Tuesday. We started in this by observations (of positions of the plane) and ended with another such observation. In between, these observations were translated into the language of Mathematics (as points on a graph), incorporated into a mathematical theory (a curve that is supposed to hold in general), and the last observation was predicted (from the terminal point on the graph).

This example is typical of the interplay between facts and theories, except that the mathematical propositions are generally much more complicated, and the connection between theory and facts is quite a bit less direct. With these few remarks as to the two "worlds" of Science, we are prepared to discuss the passage from stage to stage in detail.

INDUCTION

Induction is the process by which the scientist forms a theory to explain the observed facts. Two steps can be distinguished within this procedure: the formation of possible theories and the selection of one of these.

Let us start with the second problem. Given a large number of possible theories, how do we select the one we want? Let me introduce the term "hypothesis" to stand for an interpreted mathematical proposition which we are considering as a possible theory; I will call such a proposition a hypothesis while it is still highly in doubt, and a theory when we have accepted it. Given various hypotheses, we must first of all see whether they explain all the known facts. This is not as simple as it sounds, because our facts are never perfectly accurate, and we must face all the problems of the theory of errors (see the last chapter). But we can select those hypotheses which are reasonably well in agreement with the evidence. Then, of the remaining ones we select the simplest hypothesis.

The question that arises is: Must there be several remaining ones? Isn't it true that if we have enough facts, then there is but one theory that could explain all of these? I would like to convince you that no matter what pains you take to accumulate facts, there will always be many possible hypotheses left; as a matter of fact, there will still be an infinite number of possibilities. The best way to think of this is to think of facts as represented by points on a piece of graph paper, and of hypotheses as curves. For the hypothesis to explain all the facts, the curve must go through all the points (or, since the facts are only approximate, it suffices that the curve should pass very near all the points). Now, no matter how hard you work, you will have only a finite number of points, since within a limited existence you can accumulate only a finite number of facts. Put down a number of points on the graph, and try to draw curves through them. You will soon see that there are an infinite number of possibilities.

Of course, it may happen that of the hypotheses that *you* started with all but one will be eliminated, or even that all of these will be eliminated (since you never really consider *all* mathematical propositions as hypotheses). But there is no reason to expect this to be the normal occurrence. So you must choose from several hypotheses, all of which fit the facts. Why choose the simplest one? For the moment I will just say that it is as good a choice as any, and more convenient than the complex hypotheses. Actually there are better reasons for this, but these will have to wait for the next chapter.

In the example of drawing a curve through given points, this means that we draw as simple or smooth a curve going near the various points as possible. For example, in the airplane example this was a straight line. A straight line is always the simplest, only it is not always a possibility. As a matter of fact, scientists are so fond of straight lines that there have been many examples where a scientist has drawn a straight line through points where these points were nowhere near the line. Of course, this is a violation of the scientific method: first the hypothesis must fit the facts; only then can we worry about its simplicity.

Let us return to the discovery of the planet Neptune. What were the competing hypotheses? First of all, we could have tried some modification of Newton's laws. For example, instead of assuming that the force of gravitation always decreases with the square of the distance, we could have modified this rule. The danger in this is that the rule worked so well for the other planets. Nevertheless, we could have said that this was only because they were pretty near the sun, and hence the deviation did not show up until we got to the outermost planet, Uranus. We could have modified the square of the distance rule by a small term, which was negligible until we reached Uranus. I am quite sure that with sufficient mathematical ingenuity this could have been done, but the resulting rule would have been highly complicated, and it was simpler to formulate the hypothesis that there was an unknown planet. Of course, we run into difficulties trying to say just when one hypothesis is simpler than another. How complicated must the rule of gravitation become before we decide that it is simpler to look for another planet? I don't know. But fortunately, in most cases, one hypothesis is much simpler than the others. In our present example the rule would have become terribly complicated, so there was no doubt as to which was simplest. The only question was: Can we explain the deviations by stipulating the existence of a new planet (of the right size and in the right place)? When Leverrier showed that we could, this became the simplest hypothesis and was generally accepted even before the planet was actually sighted. We now believe in some sub-atomic particles, not because we have "seen" them (even indirectly), but because assuming their existence is the simplest hypothesis to explain the observed facts.

But how do we form the many different hypotheses from which we are to choose? To this there is no simple answer, since it is essentially a creative process. As soon as someone told us that a new planet could explain why Uranus misbehaved, it seemed most plausible. But how many of us would have thought of this possibility in the first place? How many of us would have thought that the motion of the moon and the falling of the apple are connected? How many of us would have thought that it takes no force to keep things moving, only to start them and stop them? How many of us would have guessed that the blood circulates in our veins? To select one of many hypotheses (once the facts are given) is a mechanical, even if lengthy, procedure; to think these hypotheses up in the first place is the work of genius.

There is one point on which I may be misleading you. When I emphasize the difficulty of thinking up hypotheses, you may get the impression that we must therefore choose from a very small number of possibilities. This is not so. One original idea may give rise to an infinite number of hypotheses. For example, when the idea of a new planet arises, there are the infinite number of different places where it may be (paths that it might follow), and an infinite number of sizes it may have. Let us just consider the distance of the new planet from Uranus, and its size; this already gives a double infinity of possibilities. Of course these are narrowed down by the facts, but there is still some choice: the farther away it is, the larger it must be to account for the deviations. So we start with an infinity of hypotheses, and it takes the most intricate mathematical argument to find just the right distance and just the right size to account for the path of Uranus. You may be interested in knowing that since Leverrier did not know some of the methods now available for this problem, his predictions were actually off by quite a bit.

So we see that the scientist actually thinks up infinitely many hypotheses (thanks to Mathematics), then notes which of these account for all known facts, and finally accepts the simplest remaining hypothesis as his theory.

DEDUCTION

The key to the verification of theories is that you never verify them. What you do verify are logical consequences of the theory. Verification is the process of seeing whether something predicted is really so. Since we can only observe particular facts, we must verify particular consequences of a theory, not the general theory itself.

In the case of Neptune, we could verify that there was a faint "star" at a certain location, and that the same "star" was at a somewhat different location two weeks later. But we could not verify directly how far it was from Uranus, what its path was, nor how large it was. We had to deduce some particular facts from the theory, which could be checked by direct observations.

In the second chapter we noted that logical deduction is no more than the analysis of the meaning of the theory. When we say that these facts follow, we mean that their truth is contained in the truth of the theory, even though we may not have realized this at the time we asserted the theory. When we assert that the sun rises every day, we understand that this implies its rising tomorrow. But few people, if any, would be able to look briefly at the General Theory of Relativity and see that the bending of light rays follows from it. You may feel that this is due to the fact that the theory is so complex. Then let us take Newton's very simple theory and see if it is obvious that planets move in ellipses. Or is it obvious from the theory (together with some positions of the planet) where a certain planet must be tomorrow? Certainly not, it takes long chains of mathematical deductions to arrive at these conclusions.

I have stated that infinitely many facts are contained in a theory. But it often takes intricate mathematical analysis to bring these out. So the deductive step is designed to derive observable facts from the general theories. The theoretician starts with known facts and with the accepted theories, and finds out just what follows from them. If the theories are true, then every single statement that follows from them must be true! This gives us an unlimited wealth of facts which we can check, no end to the number of verified facts which we can accumulate in support of a given theory.

The only difficulty is that the interesting results are hardly ever the consequences of a single theory, but generally of a large number of theories. So even if the prediction turns out to be false, we are not certain which of the theories is wrong. We are, however, certain that *some* theory is incorrect. It is then again a question of finding the simplest way of improving our body of theories. In the Neptune example, we altered the theory as to the number of planets, but we could have altered Newton's law of gravitation, or even his law of motion, or the laws of optics as they apply to telescopes. We changed that which was simplest to change, but we can never be certain which theory was false. At the price of making the rest of the theories sufficiently complicated, we could rescue any theory. This is the reason why we often hear the claim that each experiment tests our entire body of knowledge.

Consider an everyday example. We have been hearing a great deal about flying saucers. Perhaps by the time you read this book, the mystery will be solved. The hypothesis has been advanced that they are missiles from outer space. What I want to show is that if I am determined to maintain that the saucers have an earthly cause, no amount of evidence can shake my belief. At the moment I can maintain that they are nothing but mass hallucinations. Recently they were spotted on radar. I could try to attribute this to hallucination on the part of the radar operator. If there are too many people who see the image on the screen, I could invent an electronic effect, say, caused by too many television senders, which produce both the "saucers" and the radar images. Of course, this is likely to contradict what we know about electronics, but if I am willing to modify enough theories, I can change the electromagnetic theory, and save my pet hypothesis. If such a missile is actually shot down, I would have to abandon the hypothesis that it is a hallucination, but I could claim that it came from another country on the earth. If it turns out that there is a living being inside the "saucer," different from all we know, I could stipulate that he came from an unexplored island, or even from below the surface of the earth. Of course, to allow for life at the high temperatures below the earth, I would have to modify several theories, but if I am willing to do this, I can still save my pet hypothesis. If one of these "Martians" takes me into the plane and carries me into outer space, I can say that the machine simply took me on a rocket trip and showed me a movie which looked as if I were really looking at the earth disappearing in the distance. Even if we landed on Mars, I could explain this by the great hypnotic power he has over me.

If you got impatient with my skepticism, it was because there comes a point where accepting the fact that we have interplanetary travelers becomes simpler than modifying fundamental theories. But I hope I have convinced you that it is logically possible to save any given theory by giving up others. The reason for this is the following: In checking a theory, we must derive a consequence of the theory, which can be verified by observations. These consequences must make use of several theories, and if they do not check with experience, it is a question whether it is our theory or one of the others that is wrong. We can always suppose the latter, as was shown in the flying-saucer example. This is why the predictions are based on our entire body of knowledge, and why it is best to say that we test this whole body, rather than a particular theory.

VERIFICATION

This third step of the scientific method is similar to the first one: we gather facts. In this case, however, the facts to be observed were predicted, and we "just" see whether they are so or not.

I have tried to show that it is an oversimplification to say that one unfavorable observation can disprove a theory. This is an oversimplification for two reasons: First, the observations and the predictions are only approximate, so that we can make only probability statements (see the last chapter). Secondly, the predictions are based on several theories, and hence there is a choice as to which theory to reject (as was shown in the previous sections). So an unfavorable observation can only make the theory unlikely, or rather it makes the body of theories as a whole unlikely (more precisely, it is unlikely that the whole body is true).

How about favorable observations? First of all, these too are only approximate, so that we are never certain that the prediction was verified. Secondly, the fact that one prediction, or any limited number of predictions, has been verified does not make the theory certain. There always remain an infinity of competing hypotheses, all of which can explain all the known facts. So in this case too we can make only probability statements.

The probabilities that I have discussed here are of the second kind, credibilities. The process of verification consists in checking predictions against observations, and assigning greater or lesser credibility to our body of theories on the basis of the outcome. If the credibility is high, we are satisfied. If it is below a reasonable level, we modify our theories; we must change at least one theory so that our total body will then have a higher credibility. This may mean a relatively minor change, like admitting a new planet, or it may mean replacing the entire structure, as Relativity Theory replaced Newton's System. The decision as to when a change is necessary is complicated, and, in the absence of a good measure of credibility, is highly controversial. It is further confounded by the necessity of taking the simplicity of our theories into account. We consider the credibility of our theories and of competing ones. We abandon our theories either if some other body is much more credible on the given evidence, or if a simpler body can be found which is roughly as credible as ours.

A most interesting illustration of this point is Henri Poincaré's claim that Euclidean Geometry would never be abandoned. Poincaré was an excellent mathematician and philosopher of science. He certainly understood that there was a fundamental difference between pure, abstract Mathematics, and the interpreted version of the same, which is a branch of Science. He also understood the need for selecting the simplest possible theory. But he showed, by an ingenious argument, that one can always salvage Euclidean Geometry (properly interpreted), no matter what facts Physics reveals. He then argued that this Geometry is so much simpler than its competitors that scientists will always stick to this Geometry, and modify their other laws (of Physics) if necessary. I will just give you one example of how this can be done. It seems that there is such a fundamental difference between the finite universe of the Geometry we now favor, and the infinite universe of Euclidean Geometry, that this point should be decidable by experiment. If a ray of light can come back to its origin, then the universe is finite and closed; if it cannot, then it is infinite and open. But we can get around either possibility by modifying our other laws. Suppose a ray of light comes back after a long time (for example, we recognize, somehow, our own gallaxy in the far distance of the universe), this could be explained by stipulating that light travels not in a straight line, but in a very large circle; a circle that is so large that small portions of it seem straight. Then light could come back even in an infinite universe.

If light cannot come back, we can still hang on to our finite, closed universe. We stipulate that the universe expands (as we do stipulate) and that it expands just rapidly enough that light can get closer and closer to our "antipode" (the opposite point of the universe), but can never quite reach it, since it is running away from us with the speed of light. Indeed, some cosmological theories point to this possibility. Then light cannot return even in a finite universe. The latter case is interesting from another standpoint. If in this case we measure lengths as usual, with rulers or their equivalents, the universe is finite. But if we measure distances by the amount of time it takes light to reach from one end to the other, then the universe becomes infinite—since light can never reach the opposite "pole." Hence we can salvage one or the other Geometry, depending on the way we interpret distances.

Poincaré drew the conclusion that for reasons of simplicity Science will always keep Euclidean Geometry. He published his beliefs in a book that appeared in 1904. In 1905 Einstein's first installment of Relativity Theory appeared, a later installment of which caused us to abandon Euclidean Geometry. The dates here are quite ironic, but they do not mean that so great a thinker as Poincaré made a very bad mistake. What he overlooked was that saving the simplest Geometry might be achievable only at the price of a terrible complication in the other theories. The criterion of simplicity must be applied to our entire body of knowledge. When Einstein found that the theory explaining all the known facts (the Special Theory of Relativity) could be considerably simplified by adopting a non-Euclidean Geometry, he did not hesitate to do so. Thus the General Theory of Relativity was born. Newton's theory was abandoned because of its lack of agreement between predictions and observations. The Special Theory was abandoned because there was a simpler theory explaining the same facts. In one case, the credibility became too low; in the other case, the credibility was high enough, but there was a competing theory with about the same credibility, which was much simpler. This brings out clearly the interplay of credibility (probability) and of simplicity in the verification and consequent acceptance or rejection of theories.

A CASE HISTORY

For the concluding illustration of an application of this method, I turn to one of the greatest masters of the scientific method in history, Mr. Sherlock Holmes.

The one regrettable fact about our record of his activities is that we know him only through the eyes of the somewhat imperfect, if most likable, Dr. Watson. Among other shortcomings we find that the good doctor had his terminology twisted as far as the scientific method is concerned. He has the annoying habit of referring to Mr. Holmes' remarkable inductions (forming of far-reaching theories on scant evidence) as deductions, and of describing the scientific method—so nobly practiced by the immortal master—as the science of deduction. But never mind, let us take one of the fascinating cases and forget the terminology.

Almost any one of the adventures would serve as an illustration. I have, quite arbitrarily, selected the case of "The Red-headed League."

First of all Mr. Holmes collects facts, in this case from the narrative of a Mr. Wilson. It seems that Mr. Wilson's attention was drawn by his assistant to a strange advertisement calling for a red-headed person to collect a fairly nice salary for nominal work. Since Mr. Wilson's pawnshop is not doing well, and since he has a fine head of flaming red hair, he jumps at the opportunity. Although he is one of some thousand applicants, he is fortunate enough to get the job. It turns out that an eccentric English-born American millionaire has left a provision in his will providing for fellow redheads. All that Mr. Wilson has to do is to copy the *Encyclopaedia Brittannica,* only he must do this in an office specifically provided for this purpose. He does this, and collects a handsome fee supplementing his income, until one day (some eight weeks later) he finds the office closed and can find no trace of his employer. During Mr. Holmes' interrogation, Mr. Wilson states that his pawn-shop assistant is a most intelligent young man who has agreed to work for half-pay in order to learn the trade. The only unusual trait of his assistant is his love for photography, which causes him to spend a great deal of time in the cellar developing pictures.

Mr. Holmes has to explain the motivation for the strange employment, and also one or two peculiarities in the assistant. He formulates the hypothesis that the motives are criminal, and that the assistant has a hand in whatever crime is planned. The only purpose the job accomplished was to have Mr. Wilson out of the house; not only was he not defrauded, he actually gained a small sum. Yet there is nothing valuable in the house. The theory formed is that the part of the house which interests the assistant is the cellar (photography being an excuse only) and that the criminals want to dig a tunnel from the cellar in Mr. Wilson's absence. The fact that he was "fired" suggests that the tunnel is complete, and the fact that the crime has not yet been committed suggests that it will take place in the immediate future.

Mr. Holmes is now in the position of making some of his remarkable predictions. He can predict that there never was a will (which is verified), that there must be an important building easily accessible from the pawnshop's cellar (it turns out to be a bank), and that a robbery is to be attempted in the next few days. He further formulates the theory that it will be Saturday night, to give the robbers extra time to escape before detection. He has all his theories verified by catching the criminals just as they break in through the cellar of the bank.

This is an admirable application of the scientific method to the science of detection. A careful accumulation of facts is followed by the formation of ingenious theories. From these facts logical conclusions are drawn, which are verified one by one, until Mr. Holmes is certain of his theory (or as certain as a human can ever be). Then he can safely predict one more event, this prediction leading to the dramatic climax of the case. To those of you who still feel that there is something miraculous in the scientific method, I will give the master's own answer: "Elementary, dear Watson!"

SUGGESTED READING

Complete references will be found in the Bibliography at the end of the book.

The scientific method.
  Cohen and Nagel, Chapter XX.
  Lenzen.
  Northrop.
  Frank [2], Chapters 2, 3.

Scientific theories.
  Campbell, Chapters III-V.

Induction.
  Cohen and Nagel, Chapter XIV.
  Russell, Part Six.

Deduction.
Black, Chapter II.

Verification.
Duhem [1].
Margenau, Chapter VI.

Case histories.
Conant.
Doyle.

6

Credibility and Induction

"This conversation is going on a little too fast: let's go back to the last remark but one."

LET US RETURN to the problem raised in the last chapter but one. I have tried to show how fundamental the concept of *credibility* is to the problems of induction and to the whole scientific method. We must now consider this problem in greater detail.

THE PROBLEM OF EXPLICATION

There is no doubt that scientists assign probabilities to theories. They will say that one theory is very probably true, while another is very poorly confirmed and hence not so likely. They will consider two or more alternate hypotheses, and decide which is most likely to be true (on the given evidence). But they have no way of computing these probabilities and there are often considerable arguments as to which of two theories is more probable.

We face the same problem in everyday life. Two racing fans will consult the same dope-sheet, and arrive at different conclusions as to the likelihood of Double Negation winning the fifth race. The first decides that the horse is almost sure to win, and hence bets his shirt, while the latter only assigns an even chance to him, and hence bets on a horse giving longer odds. Double Negation wins by two lengths. Which man was right? The first man has a lot of money to back up his claim, but he has no way of showing that the odds were not even.

We are confronted with a concept about which all of us have an intuitive notion, but no one has succeeded in making the concept numerically precise. What are we to do? This problem is known as the problem of *explication,* making an intuitive concept precise.

There are various principles which must guide us in the explication of an intuitive concept. First of all, the new definition must be precise, since this is the objective of the explication. Secondly, it must agree with the intuitive concept. But this is difficult to judge, since the intuitive concept is generally very vague. At least we can require that whenever there is a clear application of the concept, then the new term must apply as well. But this still leaves a great deal of choice, and we are likely to have several precise concepts at our disposal. In that case we choose between them according to the fruitfulness of the concepts and their simplicity. The procedure is to put down clearly all those cases in which our intuition serves as a guide, and additional conditions for the concept to be useful, and then *select the simplest precise concept satisfying all these conditions.*

Let us take the concept of "hot" as our example. We intuitively classify objects as hot or cold or lukewarm, etc. While in many cases we would agree that an object is hot, in some we do not agree at all. Children differ a good deal as to what the temperature of the water must be if they are to swim in it; some will say that the lake is "very cold" while others will say (with teeth chattering), "why, it's lukewarm." The British concept of what constitutes *cold* beer is entirely different from ours. An inhabitant of Maine may consider the weather intolerably hot, when a Floridan thinks it is not particularly warm. And there are many more examples. Surely it would seem that no one concept, exact or vague, could unite these divergent views. Yet we arrived at the concept of temperature.

It was noticed that on days when most people agreed that it was warm, mercury expanded, while on cold days it contracted. So the degree of expansion of a given volume of mercury was used to construct a scale which will reproduce our intuitive judgments of "hot" and "cold." There is a great deal of arbitrariness in the construction of such a scale. What should larger numbers represent, "more hot" or "more cold"? Since the volume of mercury *increases* with heat, larger numbers have come to represent the former. But this is entirely arbitrary. What shall we choose as 0 degrees? In two scales this was taken as the freezing point of water, while in our scale the same point is 32 degrees. How shall we choose our units? In one scale it was taken by using the difference between the freezing and boiling point of water and, since this was too large for a unit, dividing it into 100 parts. In another system it was divided into only 80 parts, while in ours it is divided into 180 parts, just why I never have been able to find out. This gives us the scales in degrees centigrade, Reaumur, and Fahrenheit, respectively. Of course there are many other measures possible. Why divide the scale into 100 even parts, why not use larger units in the beginning, smaller ones later? And why use mercury rather than some other material? The reason is that this concept of temperature turned out to be very fruitful in the formation of simple theories in Physics, as any Physics textbook will bear out.

Actually it was later learned that there is a natural zero point for temperature scales. It was learned that the scale is good, because other materials also expand somehow proportionately to mercury, and hence we get simple laws as to the expansion of materials (especially gases) with an increase of temperature. Then all this was connected up with the motion of the molecules, establishing a proportion between their kinetic energy and temperature, hence showing that heat depended on the speed with which molecules moved. Thus the natural zero point is the temperature at which molecules do not move, the lowest possible temperature. The Kelvin or absolute scale differs from the centigrade scale only in having this as its zero point, thus 0° K is about − 273° C.

There are many other ways of choosing the concept of temperature, but none more useful. We could measure the heat of the summer by the number of heat deaths in the United States, but this does not lead to simple, interesting laws. We can explain the reasons for this in terms of temperature. The number of heat deaths depends not only on the temperature but on the humidity, the population density, and a number of accidental factors, such as the availability of drinking water. So we find that our scales lead to the simplest and most useful concept of temperature. But do they agree with ordinary usage? To a certain extent they must, since we chose the expansion of mercury as a measure just because it did correspond to the ordinary uses of "hot." But when two people disagree, it cannot agree with both of them. Roughly speaking, our explication says that "hot" means a high temperature, and "cold" means a low one, but just what "high" and "low" mean depends on the circumstances. In going swimming, we compare the temperature of the water, mentally, with a standard we chose. Two children may differ as to whether they like to swim in water of 60 degrees or 70 degrees, but each will use "cold" consistently to mean a temperature several degrees below the temperature he likes. Thus water of 65 degrees may seem lukewarm to one and cold to the other. Similar remarks would clarify the examples of the weather and of beer, in each case the usage of "warm" depending on what one is used to.

But the great merit of explications is that they serve to clarify such disputes, and even to get us to admit that we have made a mistake. It is a common experience that sticking a finger into lukewarm water right after we had it in hot water will make the water appear cold. In such cases we can convince the victim that he is wrong by appealing to an absolute scale, namely the thermometer. Or we might think that the weather is hot, but a brief look at our various instruments might convince us that it is not so hot, only that it is humid and there is no wind.

Even the new concept may need further clarification. It was found that the good-old 3-minute-egg takes 5 minutes on top of a mountain. But are we to admit that we like a 3-minute-egg in the valley and a 5-minute-egg on the mountain? Rather than admit such fickleness, we decided to change our scale. We called on atmospheric pressure to save face; the boiling point of water is to be taken as a standard only under normal atmospheric pressure; under the lower pressure on mountains we must assign to this a lower temperature. Actually our mercury thermometer will tell us this, but this leaves us with a choice. If we fix the thermometer so that it will show 212° F when water boils in the valley, and if it shows less when water boils on the mountain, should we say that the water boiled at a lower temperature, or that mercury expands less under low pressure? We chose the former not only to explain the dilemma of the eggs, but also because it simplifies the expression of our theories.

The problem of explicating "probability" in the sense of credibility is quite analogous. We must find those cases in which there is clear use of credibility. For example, scientists agree that the more experiments confirm a theory, the more credible it is. We must then perform the very difficult task of finding a numerical measure which will agree with all these uses, and is precise, useful, and as simple as possible. We have indicated that this work has been undertaken by Carnap and others, but it is far from finished.

Yet one thing is certain; the new concept will not agree with all uses of "probability," and even the two concepts of probability (see Chapter 4) together will not exhaust all uses, any more than "temperature" will exhaust all uses of "hot." For example, there will always remain colloquial or idiomatic uses. A colleague of mine called my attention to the idiom, "I will probably come to see you tomorrow," which is not a probability statement at all, but an idiom for, "I will have to make up my mind whether to come see you tomorrow; while I have not definitely made up my mind, at the moment I favor the decision to come." Analogously, it seems that the concept of temperature is not really referred to when a dice player exclaims: "I feel hot tonight." But it would suffice if the two concepts, frequency and credibility, exhausted all technical uses.

THE DAILY USE OF "CREDIBILITY"

In common, daily decisions we are forced to make estimates of the credibility of various hypotheses. Carnap has even suggested a method for discovering how each of us calculates his credibilities. Betting is a good example. Suppose we are offered a bet that our favorite team will win the Big Game. We are not willing to accept the 3:1 odds offered, or even the 4:1 odds, but we do accept 5:1 odds. Then we must have estimated that the credibility of the hypothesis "our team will win" is somewhere between 1/6 and 1/5. For example, we might have estimated it to be 0.18; then the credibility of the hypothesis that we will not win is 0.82 (the sum, 1, representing certainty), which is more than 4 times as great but less than 5 times as much.

Gambling is not the only possible example, however. Suppose we have to invest our money either in Government bonds or in common stocks. The stocks bring in more interest, but the bonds are safer in case of a depression. Then the amount we invest in each is a measure of the credibility we assign to the hypothesis that there will be a deflation soon.

The commonest examples given to discuss credibility involve money. The reason for this is that here it is simple to get a numerical measure of the various possible outcomes. But even where no sums of money are involved, we must make use of credibility estimates. Suppose that a student, who has completed four years in college as a mathematics major, wants to decide whether to enter industry at that stage or go on to graduate school to prepare himself as a research mathematician. Let us oversimplify this problem and say that there are just two conflicting factors. On the one hand, he wishes to become a research mathematician. We will assign a value of + 100 to the fulfillment of this desire. On the other hand, he will lose the income he would have had from his industrial position during the three years of graduate work. Let us evaluate this loss at − 30. At this point the student must make a guess as to the credibility of the hypothesis that he will be able to complete work for a Ph.D. If he estimates that his chances are about even, he will go on to graduate work. If he thinks that the odds are 10 to 1 against him, he will not go on. Of course this problem is not nearly so simple. Aside from many other considerations, there is still the question of the value put on the two alternatives. Presumably most men who go on to Ph.D.'s put a value on it that is relatively much greater than the one here indicated. And men, who are qualified to go on to a Ph.D. but do not, must place greater value on immediate financial considerations.

The degree of credibility, whose explication we have discussed, is an absolute standard to which our personal beliefs can be compared. It is the degree of belief of a "perfectly rational being" who has precisely as much information as we do. The closer our estimates correspond to the actual credibility of hypotheses, the more rational our decisions are. There are two types of reasons why someone might decide not to go on to graduate work. He may overestimate the chances of failure or he may place too high a value on the money earned during three years. In the former case, his decision would be irrational, and we may speak of an inferiority complex. In the latter case, we may describe the man as a materialist, but his decision would be perfectly rational. Again I must emphasize that this analysis was simplified intentionally; for the sake of the example I have ignored hundreds of other relevant factors.

Carnap's analysis shows an additional complicating factor, even where money is involved. It is a well-established fact that 2 million dollars are not worth twice as much as 1 million dollars to anyone. Suppose we are contemplating a bank robbery. Our local bank has 100,000 dollars that we could steal, the risk being that we might be caught and jailed. Suppose we assign to our freedom a monetary value, say 1 million dollars. Then if the odds are better than 10:1 in favor of our not being caught, we will rob the bank. But we may well pass up a bank with 10 million dollars in it if the odds are 10:1 against us, the reason being that the larger sum is not worth 100 times as much to us as the smaller one. One hundred thousand dollars may mean the satisfaction of many dreams, probably most of our dreams. With that money one could satisfy all one's desires, except perhaps the founding of a University. The 10 million dollars would satisfy the remaining dream, but that is certainly not enough to take 100 times greater chances.

This leads to the concept of the "utility" of various goals. The utility is a combination of the monetary value, how much that money means to us, and nonmonetary values, including nonmaterial ones like freedom. In every decision we must estimate the utility of the various goals, and estimate the credibility that a certain course of action will achieve them, before we can make a rational choice.

Our present discussion suffices to show that "credibility" is a concept we are in some way all familiar with, and that it plays an all-important role in our daily decisions. We must now see what role it plays in induction.

RULES OF INDUCTION

We have considered "credibility" in itself, and must now do the same for "induction," before we can bring the two concepts together.

Induction is the process of forming theories on the evidence of our observations. It has sometimes been said that induction is the opposite of deduction; deduction takes us from the general to the particular, while induction carries us from the particular to the general. The typical example given in support of this belief proceeds as follows: From "the sun rises every day" we can deduce that it will rise today, tomorrow, etc. Thus deduction carries us from the general proposition to particular ones. On the other hand, if we observe that the sun rises today, tomorrow, etc., then we form the theory that it will rise every day. Hence, induction carries us from the particular to the general.

Undeniably this sort of process plays an important role in Science. Induction, the first step in our cycle, starts with observations which are particular facts, and leads to theories which are usually generalized propositions. Then deduction, the second step, will give us certain observable particular facts which are consequences of our general theory. Yet it is misleading to say that one *always* leads from particular to general, while the other does the opposite. A theory need not be a generalized statement. A political scientist, after careful observations of the Soviet Union, may form the theory that they will attack Iran on September 1, 1965. This is a particular fact, and yet it has the status of a theory until it is proved or disproved. And deduction need not start with a generalization either. For example, from "there are at least five students in the course" and from "there are at most seven students in the course" we can deduce that "there are either 5, or 6, or 7 students in the course."

We have already formed a complete picture of the nature of deduction. It finds out certain facts which are contained in our statements, and adds nothing new (except in so far as this fact may be *psychologically* new to us; that is, we did not realize that we were in possession of this fact). This procedure often starts with general propositions, but that is not essential to its success. Induction, on the other hand, leads us from known facts to unknown ones, which we call theories. In this it may end up with a general proposition, but that is incidental to its success.

From the description of induction it is clear that it is a much more useful sort of thing than deduction. Induction tells us things we did not know before, whereas deduction only tells us things we knew already but did not realize that we knew. However, induction now sounds more like magic than like Science.

Perhaps the oldest dream of scientific philosophers is to find rules which will assure us of success in induction. These rules are supposed to lead to theories the same way that deductions lead to consequences, and they are supposed to make the resulting theories certain. The British philosophers, F. Bacon and J. S. Mill, have done most in the search for such rules, the latter actually arriving at four rules which together should guarantee success.

Let us examine one such rule. We are given a certain phenomenon, and we want to know what caused it. We consider various possible causes, and observe them in many instances of the phenomenon. Then Mill's first "canon" tells us that if these various instances have only one of the possible causes in common, then it is the cause of the phenomenon. For example, we want to find out why a certain person catches cold so often. We observe him carefully in a dozen instances where he catches cold, say by quizzing him in each case as to what he did the day before. As possible causes we consider what he ate, the clothing he wore, the condition of the room he slept in, the number of blankets he used, and his activities on the day before. We find that there was no single item of food that he ate in each case, no single type of clothing that was common to all instances, that sometimes he used two blankets, sometimes none, and that his activities of the previous day varied. But in each case all the windows of the room were open. So we conclude that it was a draft that caused his cold. Thus the rule helped us to find the cause, and since all other causes were eliminated as not possible, it proves our theory as well.

But the situation is not so simple, as Mill realized to a certain extent himself. All kinds of things could go wrong. The actual cause may be one we never thought of, such as infection. Or it may be that half the colds were caused by having no blanket on his bed, while the others were caused by walking in the rain the day before. So the rule was right in saying that neither is The Cause, but the two together did cause the colds. Or it may be a combination of causes: In some cases it was walking in the rain and not drinking anything hot after it, and in others it was a cold night and not having any blanket on the bed.

Some of these shortcomings are overcome by the other rules, but they do not change the difficulties fundamentally. In any given situation there are an infinite number of factors that may be causes. The rules will eliminate some of these, but we will still be left with an infinite number of possibilities. The rules are useful only if we have already made a guess as to what factors or combinations of factors could have caused the event, and then use the rules to eliminate some of these factors. In other words, the real merit of these rules is to eliminate possible theories, not to prove the truth of a given theory. This removes the magic element from them; we already know that Science can establish that some general theories are definitely false, though it cannot prove them to be definitely true.

Actually these rules presuppose an oversimplified analysis of the nature of scientific theories. If you look through any science book, you will find hardly any theories expressed in terms of causes and effects. Such theories are interesting only in the early stages of Science. The law of gravitation states that there is a force, proportional to the masses and inversely proportional to the square of the distance, acting between two bodies. What is the cause and what is the effect? In its useful form the law will lead to a description of how a planet moves around the sun, or rather how each moves around the other. The planet and the sun occupy certain positions which are determined by their previous positions, but are we to say that the positions at this moment are the cause of the following positions? We might, of course, call the "gravitational force" the cause, but this is a fictitious concept (as we will see in the next chapter), which is not an observable cause in Mill's sense. Such theories could never have been established by Mill's canons.

The fairly generally accepted view concerning induction is that we "posit" a hypothesis, that is, just assume it for the moment, and then try to disprove it by experiments. We accept it until disproved and then replace it by a new posit. This is sometimes called the "hypothetico-deductive method." This name is supposed to signify that the theories are only hypotheses that are assumed for the time being, and we then deduce testable consequences from them. If the theories are of a sufficiently simple form, then Mill's canons are useful for the elimination of hypotheses. What Mill does is to posit several alternate hypotheses—for example, the various different possible causes of colds—and then he tries to eliminate them one by one. If all but one is eliminated, then the remaining one is the accepted theory; but we have no assurance that further experimentation will not disprove this theory as well.

Mill does not attempt to answer the really interesting question: Just how do we form our hypotheses in the first place? Newton formed the hypothesis that the gravitational force depends somehow on the two masses and on the distance between them. Actually this is an infinite number of hypotheses, since there are an endless number of ways in which the force can depend on these factors. By comparing the predictions made on these hypotheses with known observations, he could eliminate all but one form of the theory. But how did he form these alternate possibilities in the first place? I am convinced that the formation of possible theories will forever remain a job for the creative genius of the scientist. The choice may be aided by rules, but no such rules will ever replace original thinking.

POSITING AND ELIMINATING

As happens often in philosophical discussions, we first state our position roughly, and then make improvements on it. I said that Science can eliminate hypotheses. though we noted in the last chapter that due to experimental errors, even so-called disproofs only make a theory improbable, not certainly false.

The effect of continued experimentation is to make a given theory more or less probable. This probability is an example of credibility, not of the frequency concept. Let us ignore for the moment the fact that we are far from having a satisfactory concept of credibility, and let us actually assume that the explication has been accomplished to everyone's satisfaction. Then, given a theory, and given the results of our experiments, these will give a certain confirmation to the theory, which confirmation determines its credibility. This is a number between 0 and 1, with 1 expressing that the theory is certain to be true, and 0 expressing that it has been definitely disproved. When we say that scientific theories are only "probable," we mean that the value is never 0 and never 1, no matter how close we can come to these values.

Let us suppose that we have a number of alternate hypotheses concerning a certain phenomenon. Take the chemist who has to find out what the liquid in a certain container is. His hypotheses will consist of positing that the liquid is one of various fairly well-known compounds, or mixtures of them. There are a number of standard tests which he performs, each one increasing or decreasing the credibility of each hypothesis, depending on whether the result is "positive" or "negative." Quite similarly we could have taken the doctor who looks for symptoms to decide which of several known diseases is afflicting the patient. They say that the toughest liquid that can be given to a student in his analysis-examination is distilled water. All his tests give negative results, and yet he is constantly worried that he overlooked something, or that he made a mistake. He will conclude, if he has the courage of his convictions, that the liquid is very probably distilled water. He cannot be certain, because there are certain unusual, perhaps unknown, compounds which will also react negatively to the tests he used, and there is a slight chance that his test chemicals may have been impure or that some other error has slipped in. The doctor's analogous difficulty is to decide whether his patient is a hypochondriac or suffers from a rare disease.

In these cases the decision consisted in selecting the most credible of various hypotheses. At other times we can only narrow our choice down to a few remaining possibilities and assign relative credibility to them. The doctor, in recommending an operation for a tumor, may say: "There is a definite chance that it is cancerous, but it is much more likely to be a nonmalignant growth." The result of the operation may change these probabilities considerably. Certain factors may be convincing that it was not cancer, but the recurrence of similar growths later on may lead the doctor to reverse himself. At each stage he acts on the most credible hypothesis, but he must keep in mind that his theory is never certain, and that more evidence may force him to reverse his decision.

Scientific arguments stem from two sources. Scientists may disagree as to the possible theories, for example, when on some philosophical grounds a scientist refuses to entertain a given hypothesis. Such arguments are unjustifiable, since any hypothesis considered by anyone must be allowed as possible—at least until dicredited by observations. The second type of argument arises when two scientists disagree as to which hypothesis is better confirmed by the given evidence. This shows the great significance of the concept of credibility. Although it does not help in the formation of theories, it could settle disputes as to which of various hypotheses is most credible and hence which ought to be posited for the time being.

We might even argue that an accepted concept of credibility would eliminate the necessity for original thought on the part of the theoretician in Science. Why not consider all possible hypotheses and calculate which is most credible? Of course we would have to calculate the credibility of an infinite number of hypotheses, but Mathematics does have methods of finding which of an infinite number of quantities is greatest. Such maximum calculations are used in many mathematical problems. But there are two essential difficulties. Not all possible hypotheses can be formed in one language, as we have seen in the third chapter, and so the choice of a language remains creative. This does not sound much like the kind of thing a scientist does, but actually some new scientific discoveries are based on an entirely new terminology. Relativity Theory introduces the language of tensors, Quantum

Theory the language of the ψ-function, and if the Theory of Games becomes a useful tool in economics, we will have the language of strategies. Even within a given language it is very questionable whether the problem of finding the most credible hypothesis is a solvable problem in all cases. We cannot possibly decide this until we have the explication at our disposal, but it rather reminds me of certain logical problems which have been shown to be unsolvable.

We can now formulate scientific procedure as follows: On the basis of given observations the scientist formulates a number (very often infinite) of possible hypotheses. He selects that hypothesis which is most credible on the basis of the observations and posits it as his theory. He then proceeds to make predictions which he tests, each test modifying the credibility of the hypothesis and of its competitors. The hypothesis remains the posited theory until its credibility becomes less than that of a competitor, in which case it is "rejected" in favor of the alternate theory. Later experiments may, of course, eliminate the alternate as well, possibly even reinstating the original theory.

There is one more factor that we must add to our description of the positing of hypotheses. Since our possible theories can be infinite in number, it may happen that many theories have approximately the same credibility. In the case of the law of gravitation it may be clear that the force must decrease as a certain power of the distance which is approximately the second power. But we do not have enough information to make it even reasonably certain that the right power is not 2.0000001. It may even be that the latter happens to be in slightly better agreement with observed facts than the power 2. But since we know that due to experimental errors this may be caused by a minor mistake in observations, we really have no good way of deciding whether to choose 2.00000000 or 2.00000010 or something like 2.00000006. Hence we choose the simplest of these alternatives, namely, the value 2. This is certainly dictated by considerations of convenience, but I have shown elsewhere that there are good methodological reasons for choosing the simplest of the highly credible hypotheses. We have faith in the method because we hope that successive experiments will eliminate all powers far from the true value, and hence our posit will come closer and closer to the true value. But this gives no reason to suppose that we will ever find the exact truth. I have succeeded in showing that, if the true hypothesis is one of the ones under consideration—in our example, if the true exponent is expressible as a finite decimal expansion—then by always choosing the simplest credible hypothesis we will not only approach the truth, but we are almost certain to find the true value.

We can sum this up by saying that the scientist forms a number of alternate hypotheses and performs experiments which make these alternatives more or less credible. He rejects all hypotheses whose credibility is considerably lower than that of some other alternative, but instead of selecting the *most credible* hypothesis, he posits the simplest hypothesis having a high credibility.

This principle can be well illustrated in examples where the data are put on a graph paper. Suppose that we have ten points which lie approximately but not exactly on a straight line. The most credible hypothesis may be represented by a waving line which goes through all the points, but it need not be the true hypothesis. Due to minor errors we may have gotten these points even if the true law is represented by the straight line. If the points are not too far out of the straight path, then the straight-line hypothesis will be highly credible and must be selected in preference to the slightly more credible but less simple wavy line. Now let us suppose that the next observation is so far off this line that we must reject our previous posit. Then again we consider various possibilities, curves in this case, which "fit" the points fairly well, and select the simplest of these, say a parabola. (This may be considered the next simplest case, since a straight line is represented by a polynomial of the first degree, and a parabola by a polynomial of the second degree.) At each successive stage we keep our posit as long as it has a sufficiently high credibility, and when it fails, we replace it with the next simplest possibility having high credibility. Thus Science progresses by successive posits, resulting from induction, and eliminations, resulting from deductions and verifications which together modify the credibility of the various hypotheses. We never reach certainty, but our theories become more and more probable all the time.

THE JUSTIFICATION OF INDUCTION

The procedure here described has a certain intuitive plausibility, and this justifies it in one sense. But the traditional problem of justification is to *prove* that this method will be successful, or at least make it very probable. After all, we stake our lives on the predictions of Science every time we go up in a plane, take a drug, or even when we just go to bed with the assurance that the roof will not collapse.

A number of different approaches have been tried. The most famous of these is the justification of this inductive method by arguing that it worked so well in the past. This is the most famous example of a vicious circle in philosophical arguments. The inductive method rests, roughly speaking, on the assumption that what held true in the past will also hold true in the future. If we justify this assumption by saying that it worked well in the past and hence it should work in the future, then we are using the very assumption we are trying to prove.

Another modern and sophisticated approach is to say that the theories are only conventions and that experimentation shows only whether our conventions were good or bad, not whether they were true or false. But if the theory is only an arbitrary convention, then we have no reason to suppose that predictions based on the theory will be found to be true.

We have also heard arguments that induction leads only to low-level generalizations, like the one that all planets move in ellipses, which are then proved by being deduced from laws—the law of gravitation in this case. But who is to say that the law of gravitation is true? Ironically enough, Newton's law has already been disproved.

We could try to base the justification on our concept of credibility. Although the prediction is not certain, it is highly credible or probable. Isn't that enough? This turns out to be the modern form of the vicious circle argument. The very method of calculating credibilities assumes that the past is a reliable guide to the future. For example, the more often we have observed the sun rising without its failing to rise, the more credibility we assign to the hypothesis that it will rise tomorrow. Credibility bears out our inductive assumptions, because it was constructed to do precisely that.

In short, induction cannot be justified. We can only base it on a more or less plausible sounding assumption, as we have seen in the third chapter. We may argue as to the best way of forming theories, but whatever method we arrive at, we will know only that the theories agree with past observations, and we must take on faith that they will hold true in the future. Only in one sense can we justify induction. While our assumption must be taken on faith, we can point out that unless some such assumption were true human life would be impossible. If nature were designed so that plausible inductions invariably turn out to be wrong, the human race would be wiped out soon. We may not be able to justify the assumption, but we must have faith in some such assumption if life is to be possible!

SUGGESTED READING

Complete references will be found in the Bibliography at the end of the book.

Explication.
  Carnap [2], pp. 3-18.
  Hempel [1], pp. 6-14.

Rules of induction.
  Mill [1], Book III, Chapter VIII.
  Cohen and Nagel, Chapter XIII.

Justification of induction.
  Hume, Section VII.
  *Reichenbach [2].

Utility.
  *Carnap [2], pp. 264-279.

Simplicity of theories.
  *Kemeny [2].

7

Concepts of Science

> *"What's the use of their having names," the Gnat said, "if they won't answer to them?"*
>
> *"No use to* them," *said Alice; "but it's useful to the people that name them, I suppose. If not, why do things have names at all?"*

IN THE LAST CHAPTER we discussed the formation of theories. Nothing is more obvious than that theories need words for their expression, but the very fact that this is obvious resulted in no attention being paid to the words. At the beginning of this century certain concepts of Science were questioned, and rejected as unhealthy. Since then the nature of scientific concepts has become a highly controversial and interesting issue.

TYPES OF CONCEPTS

It is the task of Science to record facts and to form theories to explain and predict observations. Corresponding to the distinction between observations and theories, it has suggested that we distinguish observational and theoretical terms.

Observational terms are supposed to describe things directly observable. The reason I said "supposed to" is that it is not so easy to say what is directly observable. Direct observations are those which cannot possibly be wrong, in the sense that we state only what we actually see at the moment. But by the time we put this down on paper, we usually have read something more into our observations. Suppose I report that there is a large bookcase in the room. Strictly speaking I have no right to say that. What I see is a blot of dark brown, with things sticking out at regular intervals, and it seems to occupy a space about as high as I am. So much is certain. But that this is a bookcase is an inference which may be wrong; it may be a built-in-closet, or even a painted decoration on the wall. It may not even be large. It may be a small object with a trick magnifier in front of it. Or the whole thing may be something projected from a movie projector behind me, or even a hallucination.

We have observed a certain impression of something bookcase-like, and assuming that it is neither a hallucination nor a deception, there really is something looking like that in the room. But if we say that it is a bookcase, then we infer that it will also look like a case from other angles, which we have not observed. What happened was that we had a certain sense-impression, and since such sense-impressions have usually turned out to be associated with the presence of large bookcases, we have inferred that there is a large bookcase in the room.

The usual reports of observations already contain theoretical inferences, though they are very low-level inferences, based on everyday generalizations. One of the very interesting questions in the Theory of Knowledge is just what should constitute pure observational material. According to one school, the phenomenalists, reports should be restricted to accounts of sense-impressions; the other school, that of the physicalists, holds that the basic statements are reports of the properties and relations of physical objects. Without trying to minimize the importance of this dispute, we will ignore it, because it would carry us far afield. Instead we will allow "there is a large bookcase in the room" as an observational report, keeping in mind that such reports may at times be incorrect. Our observational terms will be words like "brown," "large," "between," "solid," "sweet," "warmer than," etc. Actually "bookcase" is not one of our basic observational terms at all, but is shorthand for an observational description something like "solid object with right angles, closed on five of its six faces, with protruding horizontal sheets, etc."

When we look at the theories of Science, we find entirely different sorts of terms. Let us again take Newton's famous Law of Motion as an example: $F = m \times a$ or "The force equals the mass times the acceleration (in suitable units)." The three key terms "force," "mass," and "acceleration" all refer to quantities which are not directly observable. Let us picture the flight of a home-run ball hit by Stan Musial. By holding it in the hand we can tell its weight approximately, we see Mr. Musial exerting a force, and we note the rather considerable acceleration imparted to the ball. But we certainly cannot assign numerical values to these observations, and, as we will see later, these numerical values are determined in a very complex way.

There seems to be a logical paradox in the formation of theories. Our initial data are stated in the language of observations, while the theory formed is in an entirely different language. At first sight this is like inferring that all men are mortal from the observation that certain bees collect food from flowers. The only way the inference can be made plausible is that we supply a "translation" from the language of theories to the observational language. This is done by means of the rules of interpretation. These rules have received careful study lately, and we are presented with a very complex picture.

We can illustrate this by considering "acceleration." The rule of interpretation tells us to observe the position of the ball at several different moments. In the case of straight-line motion—which is approximately the case for a line-drive home run—we simply divide the distances traveled between successive positions by the time elapsed. This gives us approximate velocities at these moments. The rule further tells us that the more observations we make, the more accurate our estimates of the velocities will be. The exact velocities are "limits" of these approximations. Then we arrive at the acceleration at various moments by a similar process applied to the true velocities. Hence the accelerations are not determined by any finite number of observations. They are derived through a complex mathematical operation applied to an infinite number of observations (that is, we assume that we know the position at every moment and differentiate twice). The only direct connection with experience is that we can get a good approximation for the acceleration at any moment, if we are willing to make sufficiently many observations. The connection to experience of "force" and "mass" is even more indirect.

In this chapter we must approach the true picture step by step, filling in more accurate details as we proceed. For the moment we can say that we have a set of observational terms for the description of the facts given through observations and experiments, and that theories are formed partly in terms of these and partly in terms of theoretical concepts which are only indirectly connected with experience, the rules of interpretation providing the bridge. As we go to more advanced theories we find fewer observational terms, if any, and the theoretical terms become further removed from experience.

OPERATIONALISM

The revolutionary work of Einstein started with a reconsideration of the concept of time. When someone says that two events took place at the same time, everyone knows what he means. Yet Einstein showed that such statements are often, strictly speaking, meaningless.

When someone says that he lifted both hands at the same time, that has a perfectly definite meaning. But when we consider events taking place far from each other, we may find two people who will disagree as to whether they occurred at the same time, and both of them will be right! Let us attack this paradox by seeing how we would determine whether two events occurred at the same time. Suppose we note that the clock in the tower strikes just as we reach our door. Did the two events really take place at the same time? We would be tempted to say that they did, but we must remember that it takes time for sound to travel, and when we hear the clock strike, it really had struck a second or so ago. But couldn't we take this into account by allowing for the time it takes for the sound? We could in this case, but what if the sound is given off by an airplane, which is traveling itself? Let's try something different. The pilot could look at his watch when the sound is given off, and we could note the time as we enter the house, and compare later on. This will work, if the clocks are synchronized, which means that the pilot's watch shows the same time at the same moment as ours, but we are now just trying to determine what "the same moment" means.

The whole trouble arose out of the error caused by the time it took sound to travel. Can't we reduce this by using a faster method of comparing watches, and thus reduce the error as much as desired? For most of the history of Science this was a perfectly satisfactory answer, but we now know that there is no speed greater than that of light, and hence there is a limit to how small you can make this error. Fortunately, light travels so fast (for example, from the sun to the earth in 8 minutes) that the errors are negligible for practical purposes, but no error is negligible for Science if it is *impossible* to reduce it.

Let us finally try to determine whether the events are simultaneous by putting an observer in the middle. Then it does not matter how long it takes the signals, as long as they travel with the same speed. If signals are set off at both events, say flashes of light, the observer need only check whether the signals reach him at the same time. This sounds convincing on traditional grounds, but it is not a solution to the paradox. It works only if we know that the observer is not moving. Should he be moving, then—thinking traditionally—he would meet one of the signals—the one he is approaching—sooner than the other one. But motion has been shown to be relative; we can tell whether two trains are moving relative to each other, or to the earth, but absolute motion is meaningless. So we still do not know exactly whether the events took place at the same time.

The conclusion Einstein arrived at is that two events which may appear to be simultaneous to one observer may not be so as viewed by another observer. This decribes, very roughly, the argument that convinced contemporary physicists that absolute time, as conceived by generations of scientists and men-in-the-street, is an abstraction to which nothing measurable corresponds in nature.

The operationalists trace the origin of their doctrine to this argument by Einstein. The operationalist thesis is that every scientific concept must be connected to experience by means of precisely given operations, which tell us how to apply the concept. Of course, this is easy for observational terms. The problem is to furnish such "operational definitions" for the theoretical terms. For example, if a biologist is to use the term "mammal," he must be able to tell us exactly how we determine whether a given animal is a mammal.

Operationalism is often described as a modern version of Empiricism. Empiricism is the school which maintains that all knowledge must originate in experience, a belief that in its many variants is undoubtedly the most popular tenet of Modern Philosophy. Operationalists argue that all useful concepts must originate from experience as well, and hence require operational definitions for all theoretical concepts. There is no doubt that if these concepts are to describe experience they must somehow be connected with it, and in calling explicit attention to this the operationalists have performed a great service. But just what this connection should be is still open to dispute.

It is a natural tendency that when we notice something that is novel and important, we state our case in an extreme form. The operationalist requirement for explicitly statable rules as a prerequisite of the use of scientific concepts seems to be an example of this. We find that practically no fruitful concept fulfills the operationalist requirement entirely. We need only examine the discussion of "length" given by Bridgman, foremost advocate of the operationalist position. As we have already noted for "acceleration," the precise value is not gotten unless we have performed an infinite number of operations, which is not possible. So the best we can ask for is that the values be approximated as closely as desired. But even this requirement would eliminate the classical concept of time, since the existence of a maximum velocity put a limit on the possible approximation. And even much graver difficulties arise.

How would we go about measuring a length, even approximately? We would take a yardstick and put it down next to the object measured, and see how many times we can lay it end to end—forgetting for the moment about fractions of yards. But which yardstick? There is an official meterstick in Paris, but it is very precious, and never used. What we do is measure off another stick of the same length, and use it. But how do we know that this will give us the same answer? How do we know that it makes no difference what material it is made of and by what route it is carried to the object to be measured? The answer is that we consult our theories to see whether these factors influence the length; we are told that the route traveled makes no difference, but the material it is made of may influence the measurement, for example, in very hot weather. But this is a vicious circle! We are trying to define length; in applying this definition we must apply theories which make use of the concept of length, which is yet to be defined. How do we know what the theories mean until we know what length is, and how do we know what length is until we have these theories?

There are many cases where measuring rods are of no use. What yardstick will measure the diameter of an atom, or the distance between two stars? These lengths are determined by complex, indirect methods, with theories telling us that these methods will yield the same result as actually laying down a yardstick. In the case of the two stars, for example, we observe some light coming in through our telescopes, and after lengthy calculations based on the laws of optics and astronomy we arrive at a distance. It is entirely conceivable that rocket ships may some day be able to perform the same measurement with yardsticks (though this may require a few million years to perform) and we have only our theories to tell us that this result will be the same as our present estimate.

In everyday applications as well as in the most demanding laboratory we think nothing of turning yardsticks around, or of putting several sticks end to end instead of moving the first one. When we assume that this still gives the correct length, we are taking theories for granted, though in this case very highly confirmed theories. Bridgman states: "In general, we mean by any concept nothing more than a set of operations . . ." and "if we have more than one set of operations, we have more than one concept. . . ." But this would outlaw most of what scientists actually do.

The formation of concepts and of theories using these go hand and hand, and both the concepts and the theories are modified with the progress of Science, the two processes being inseparable. We formed an intuitive concept of length and formed the theories stating that measurement of length was independent of many factors, such as the material of the stick, how it is moved, whether one or more is used, which way it points, what the temperature is, whether a man or a woman does the measuring, and what the measurer's political opinions are. Soon it became apparent that the body of theories so formed must be modified; it was simplest to do this by admitting that temperature and the material of the rod influence the length, and new theories were formed to "correct" for temperature. Even more recently, 1905 to be precise—this being the historic date of the Special Theory of Relativity, we admitted that the motion of the rod can affect the measurement of length. If the pilot of a fast rocket ship should take with him the standard meter, and we tried from the ground to measure this stick in flight, we would find to our horror that it shrank. Fortunately, as soon as the flight ends it regains its standard length. Put this way it has amazed two generations of laymen, and a good many scientists as well. Actually this means only that the results of measurement depend on motion, just as they depend on the temperature. This further modifies our concept of length, and new theories go with the new concept, the two changes being inseparable.

The operationalists trace their work to Einstein's work, and yet Einstein is not an operationalist. In reply to an essay by Bridgman, Einstein summarized his position as follows: "In order to be able to consider a logical system as physical theory it is not necessary to demand that all of its assertions can be independently interpreted and 'tested' 'operationally'; *de facto* this has never yet been achieved by any theory and cannot at all be achieved. In order to be able to consider a theory as a *physical* theory it is only necessary that it implies empirically testable assertions in general." It is certainly admitted that the theories must be connected with experience, there must be predictions following from the theories which can be tested by perfectly precisely determined operations. But it is not necessary that each concept be so defined, and even that every statement be so testable.

We find theories stated mostly, if not exclusively, in theoretical terms. The rules of interpretation connect these with experience, but certainly this does not mean that the theories can be translated into observational terms. If they could be, the theoretical terms would serve only as abbreviations. The case is rather that, from the theories, by means of the rules, there follow observational statements, which can be checked against experience. Take the example of a baseball in flight. Applying Newton's law, $F = m \times a$, we get such statements as $a = 6.3$ at such and such a moment. But we have seen that this cannot be directly observed. It also follows that the average velocity will be 60 feet per second near a certain place; the rules of interpretation will tell us that this means that if we clock the flight for 3 feet on either side of this point, it will take about $1/10$ of a second. This latter statement is in observational terms and can easily be checked.

If a theory has no precisely testable consequences, then it tells us nothing about experience. But as long as it has such consequences, and all these predictions turn out to be true, it makes no difference what other statements it may make which cannot be checked. The two criteria are that the theory should make only true predictions, and that it be as simple as possible. It is often assumed that the extra paraphernalia of untestable and hence "useless" statements unnecessarily complicate the theory. But the least reflection will convince us that this is not even plausible. When the ball is in flight, we can check only a finite number of positions, and yet it is much simpler to form a theory of the entire path, not only of those points which can be checked. Of course here one could reply that any one point could be checked, even if we cannot check all. But we also know that concepts like velocity can never be measured exactly. Even concepts which can never be checked directly have proved to be so useful that Science would be sadly retarded without them.

Some early operationalists refused to accept atomic theory because the concepts used cannot be defined through operations, i.e., because atoms cannot be seen. The theory survived all such criticisms. Perhaps genes are the best example of the fruitfulness of operationally untestable concepts. Genes cannot be observed today, and it would make no difference if we could never observe genes. Genetics would be just as useful as a theory. The theory would have untestable statements, such as "This flower has one gene for 'red' and one for 'white.'" But through the rules of interpretation, predictions like "Approximately one quarter of the offspring will be dominant, one half hybrid, and one quarter recessive" can be translated into observable predictions about having roughly three quarters red flowers and one quarter white ones, and about the appearance of the following generation.

Consider a box which is locked in such a way that it is impossible to open without destroying the contents entirely. You observe the workings of the box; for example, it may perform certain calculations for you. By far the simplest theory that you can form is to try to guess what kind of machinery is in the box, even if you can never check this directly. You may form the theory that it contains an IBM machine, of a certain type, but there is no way of checking this. You can predict just how the box will act under various circumstances. For example, if you insert a card with 3 and 5 punched into certain columns, it will make clicking noises, and a card will come out with 1 and 5 punched in successive columns. As long as your predictions continue to be verified, you will maintain the theory, even though the machinery is inaccessible. But the theory that there is a ghost in the box is not scientifically acceptable, because it has no testable consequences.

FICTITIOUS CONCEPTS

We have seen that theories must be connected with experience by means of observable predictions. On the other hand, we have not agreed that all statements must be testable, or that all concepts must be operationally definable. As a matter of fact, many concepts which the operationalists accept are best described as "fictitious." "Force" is a good example of this.

Newton's famous Law of Motion states that $F = m \times a$, where F is the force, m the mass of the body to which the force is applied, and a is the acceleration measured in suitable units. We have already discussed how the acceleration can be measured; let us ignore for the moment the difficulties in the operational definition. But how are we to measure masses and forces? According to the operationalists, these two concepts must be precisely defined first before they are used in a law. We determine a mass by comparing it to an arbitrarily chosen unit mass. We apply the same force to both masses, and if the first gets twice the acceleration of the second, the second has twice the mass of the first. We measure forces by comparing them to an arbitrarily chosen unit force. We apply both to a given mass, and if the force we are measuring gives twice the acceleration to the mass that the unit force gives, then it is a force of magnitude 2. These are operational definitions, but they make implicit use of the law!

Actually, so far we have used only two special cases of the law in question. We have made use of the fact that, for a given mass, accelerations are proportional to forces; and that, for a given force, accelerations are inversely proportional to masses. So we can say that this much is a matter of definition, but $F = m \times a$ says a great deal more, since it allows the comparison of a force applied to a mass with another force applied to a different mass. But actually much deeper use is made of Newton's law. Suppose I push a billiard ball and give it a certain acceleration. Then I push a second ball having the same mass, but it is given a greater acceleration. The conclusion would certainly be that, knowingly or unknowingly, the second push was one of greater force.

It is very difficult to escape the conclusion that Newton's law is not a theory but a definition of force. It gives you an operational determination of the magnitude of a given force. It would be a genuine law if we had a way of measuring forces independently of the law; as it is, the "law" is but an agreement as to the way forces are to be measured. The situation is further complicated by the fact that masses cannot be measured without using this law, but in this case it suffices to have a fixed force (for example, the gravitational force at a given point on the surface of the earth), and in terms of this force we can give an operational definition for masses. The definition is simply a description of the process of weighing. Let us suppose that we agree to do this. We choose a unit mass arbitrarily, and determine other mases by weighing at some fixed place, say the physics laboratory of a centrally located university. Then we have operational definitions for m and a, and $F = m \times a$ serves as a definition of F.

Does this mean that the law has no factual content at all? Not quite. If one adds a certain intuitive notion of when we have applied the "same force" twice, then we can require that in each case $m \times a$ be the same. This intuitive notion undoubtedly arises out of our experience of exerting physical "force." We can roughly judge whether we pushed the two billiard balls equally hard, and any definition of force would be unacceptable if it did not bear out our feelings reasonably well. So we expect that when different objects are brought into the "same situation" the forces applying to them will be the same. There are, however, important exceptions to this, and in each case it is the idea of the forces being the same that is abandoned, not Newton's law. For example, the force (of gravitation) applying to different objects when dropped out of a window varies according to their masses. Electromagnetic forces depend on the amount of charge on a body and on the material of which it is made. It is such cases which prove conclusively that, in the last analysis, $F = m \times a$ is a definition.

Then why is this "law" useful? Because together with other laws it tells us a great deal about experience. For example, Newton's law of gravitation tells us that the force of gravitation between two bodies is proportional to their masses and inversely proportional to the square of the distance between them. If in this law we substitute $m \times a$ for the force, we get an expression for the acceleration due to gravity, which is a very important law. But this law no longer involves forces! In general, we find that every statement about actual experiences can be stated without making use of the concept of force.

In other words "force" is a fictitious concept introduced to stand for $m \times a$, brought about by our psychological experiences of exerting a force. We seem to transfer this feeling of strain to Nature, when we speak of a gravitational or electromagnetic force. It may be a useful device psychologically, but it is logically unnecessary. As the problem is formulated, F is defined by Newton's "law," and in any given instance we must find on what the force depends. We could equally well have omitted the concept of force and determined directly on what the acceleration depends. For example, in the case of a spring, we find that the force is proportional to the extension of the spring; and hence, since $F = m \times a$, the acceleration is proportional to the extension of the string and inversely proportional to the mass of the body. All observable predictions follow from this last law, and hence the concept of force was superfluous. About all we can say is that, since accelerations are frequently inversely proportional to masses, the product of these two is a convenient quantity for which to have a name.

We often use this procedure in everyday arguments. Suppose we decide to "type" human beings. This is generally based on noting that certain characteristics frequently occur together. If we had a complete theory, we would classify every human being as being of a certain type. We could then say a great deal about him according to the laws we have found applicable to his type. Of course, in all these cases all that we have to say can be said by talking about combinations of characteristics, and these "types" are no more than convenient names for certain frequent combinations. If we try then to say that the statement "type so and so always has such and such characteristics" is a law, then we confuse laws and definitions, as has happened to $F = m \times a$. It is typical of the scientific study of man that many scientists have been impressed with the progress made, when all that has happened in many cases is that a series of definitions was introduced. The test of a new concept is in the resultant theories. If we can formulate strong theories about personality in terms of our types, then they are useful fictions; otherwise they are no more than useless exercises in defining terms.

RULES OF INTERPRETATION

We have agreed that while the requirements of the operationalists are too severe, we must still require that there be definite ties between theories and observations. The rules which establish such connections are called rules of interpretation.

These rules will tell us which of the statements in our language describe observable phenomena, and just what observations will establish whether the predictions are right or wrong. The ancient oracles were made famous by making predictions and *not* supplying rules of interpretation. Since the predictions were in terms of ordinary language, they seemed to make sense. Actually they were worded so that no matter what happened, the oracle's prediction could be twisted so as to appear to have been correct. A scientific theory without definite ties to experience could also be molded to the facts after they are discovered.

A good example of such a theory is Freud's theory, not as interpreted by a competent psychoanalyst, but as popularly understood. Let us apply it to dream analysis. Every dream is caused by Sex. A man dreams that he is working in an airplane factory, that he is accused of sabotage, and he knows that he has an excellent alibi for the night of the crime, but he can't remember what it is. So far Sex has not entered. But why dream about an airplane at all? Well, when was the last time he rode in a plane? He never has. When was the last time he spent some time looking at a plane? It turns out to be at a museum. After being pressed on this point, the "patient" admits that there was a very attractive woman in the museum, and that it is possible that he did give some thought to her. It is then an easy matter to figure out that the sabotage symbolizes immoral thoughts, and that the difficulty in remembering his alibi was caused by trying to think up ways in which he could pull the wool over the watchful eyes of his wife. So Sex *is* the explanation.

There are two types of objections that immediately occur to us. First of all there are numerous alternate explanations possible. We may have seen a movie about sabotage the night before, there may have been an airplane flying overhead whose sound we heard in our sleep, and the alibi may have been an expression of guilt over not having performed an act we considered our duty. Secondly, we may object on the ground that if we just push our inquiry far enough, we can eventually bring any subject matter into it, even Sex. It is a common joke about college "bull sessions" that if a group of men talk long enough, they are bound to come around to women. It is probably just as true that if they talk long enough they are bound to come around to medieval art, but they are less likely to note this.

In the next chapter we will see that an explanation is acceptable only if it could have been used to predict the outcome. Yet I have heard of no one who claims that knowing that a certain man saw an airplane in a museum, and knowing that he paid some attention to a certain attractive young lady in the museum, and knowing that the young lady had blonde hair—while the man's wife is a redhead, will enable one to predict that the man will at some future time have a dream about sabotage in an airplane factory, and about forgetting his alibi. The same applies, of course, to the alternate explanation.

The difficulty is that we have no definite connection between "All dreams are caused by Sex" and observations. In order to establish such a connection, we would have to specify more precisely what we include under the very broad heading of "Sex," and just exactly how some sexual experience connects with a dream. This is not so difficult; but if it is done, then our theory becomes testable and may turn out to be false. Much has been done along these lines by disciples of Freud, but an evaluation of the new theories does not concern us. This is merely an illustration of the kind of ties that must be supplied with experience.

In the second chapter we discussed the connection between Mathematics and Science. We saw that pure Mathematics must be interpreted before it becomes a scientific theory. In this case the need for rules of interpretation is clearer than in the case of theories stated in words. It makes no sense to ask whether $E = mc^2$ unless we interpret E, m, and c in a manner that will connect the law with experience. The connection in this case consists of rules as to how one measures the amount of energy (E) and mass (m) present, and how one determines the velocity of light (c). Then we say that in every process in which there is a loss of mass, there will be a corresponding increase in energy, given by the theory. Thus, if in an atomic engine we succeed in disintegrating an ounce of uranium entirely, we must expect an increase of energy c^2 times that much, or an increase of energy equivalent to that which a 100-horsepower engine will produce in about a million years. This clearly connects the theory with observable facts.

The difficulty arises in that it is often not easy to separate the theory from the rules of interpretation. Theories like Freud's carry their interpretations with them in that they are formulated in common language. They also carry all the usual vaguenesses of such languages. When we pass to mathematical theories, we frequently find that the theories and rules are formed jointly. We have already tried to analyze $F = m \times a$ and have seen how hard it is to separate the factual assertions from the definitions of the terms. But while it is difficult to separate these factors, in every good theory both elements clearly exist. On the one hand, we have mathematical formulas which we manipulate by the rules of logic; on the other hand, we know that certain types of assertions express observable facts. The crucial requirement is that we be clear about these assertions and what facts they describe, so that when we make an observation in the field of the theory, there can be no doubt as to whether this agrees with the prediction of the theory or not.

We have noted in the $F = m \times a$ theory that the theory contains a rule of interpretation (for F) within it. There is the converse case as well, where a supposed rule has a theory hidden in it. As a matter of fact, this is very common; we have already noted that so-called operational definitions generally assume that various similar procedures will lead to the same result. When we define length as the result of laying down measuring rods, we assume that the result is independent of such factors as which end of the rod points North. We would like to say that rules of interpretation are entirely arbitrary, like definitions, that their justification is usefulness, not truth. This would be the case for a pure rule; but because interpretations are closely tied to theories, rules may prove to be unacceptable—for example, the classical rule for determining simultaneity. Hence we adopt the attitude of judging theories and their interpretation (connections with facts) as a single body of knowledge.

CONCEPT FORMATION

It should be clear by now that there is a great deal of arbitrariness in the formation of concepts. We tend to say that objects have masses, velocities, energy, etc. Actually these are free creations of the human mind, which have proved useful for the formation of theories about experience. Instead of using the concepts of mass and acceleration (m and a), we could invent the new terms meleration (M, standing for $m + a$), and accelass (A, standing for $a - m$). Then, using the same concept of force, the force is no longer the product of mass and acceleration, but 1/4 the difference between the squares of the meleration and the accelass; that is, $F = m \times a$ becomes $F = \frac{1}{4} (M^2 - A^2)$, as a little bit of algebra will show. Objects *have* melerations and accelasses just as much as they have masses and accelerations, but the new concepts make our theories much more complex.

The concepts of force and acceleration also serve to illustrate the way concepts change through the history of Science. We have considered these concepts so far only in connection with motion in one given direction, along a straight line. But suppose that the motion is not straight, that the direction keeps changing. Our old concepts of force and acceleration no longer apply; rather, we could still apply them as before, but they no longer lead to simple theories. However, Newton discovered that by generalizing these concepts we get laws for curvilinear motion that are as simple as those for straight lines. We think of velocity as having various components; three components in three dimensions. Then any change in any component is an acceleration. Force too has three components, causing changes in the corresponding components of velocities. Then we can keep $F = m \times a$ as our definition of force, and such laws as the law of gravitation still apply.

But this is a misleading description. If in $F = m \times a$ or in the law of gravitation we change the meaning of the terms, we no longer have the same definition and the same law. This is an interesting case where progress in laws is brought about by keeping the same form of the law, but changing the interpretation of the terms. In general, of course, we change both the concepts and the form of a law, and we change these simultaneously.

These examples also show why it is often hard to say what is the theory and what the definition. It happens so often that our body of knowledge is kept simpler if we are willing to change a definition than if we change the form of our theory. Hence, it can happen that when a scientist finds that his prediction turned out to be false, he decides to modify a definition. He may even describe this by saying that he found out that his definition was false. This means, of course, that he thinks of his definition as a theory.

Indeed, the same body of knowledge can be set up formally in several ways: according to some, a given statement would be a theory; according to another, a definition. Hence, while the distinction is important to the logician, it is of no practical importance to the scientist.

In Chapter 9, we will see that there are various levels of theories, with more and more general applicability. Since the theoretical concepts are tailor-made for theories, we find various levels of concepts. We might start with observations, which are stated in observational terms. Then we have "direct generalizations" from experience, which may still be in observational terms, or in terms of theoretical concepts having a very direct connection to experience. As we go to more general theories, the concepts are more abstract, and operational definitions become increasingly more difficult if not impossible.

Let us consider some examples. "This liquid is blue" is an observational report, and "liquid" and "blue" may be considered as observational terms. "Hydrogen combines with oxygen to form water" is a low-level generalization. Terms like "hydrogen" are not observational terms, since no observation will "tell" you that a given gas is hydrogen. However, it is easy to give an operational definition; this would consist of the usual chemical test for hydrogen. "The atoms of hydrogen have one electron" is a more abstract theory, and it is impossible to give an operational definition for "atom" and "electron"; the rules connecting this theory with experience are indirect. And when we come to Quantum Mechanics, the ψ-function is one of our "most theoretical" terms. It is difficult to give any popular description of how this theory connects with experience. The connections on this level are of the form that certain complex mathematical statements in the theory are predictions of observable facts. The general attitude one finds among the adherents of this theory is that all that matters is whether each such prediction turns out to be true and that it is nonsense to ask whether the rest of what the theory says is correct or not. No matter how we may feel about this from a metaphysical standpoint, this is certainly all that matters for practical applications.

So we are confronted with a hierarchy of more and more abstract concepts. The justification for each concept is its usefulness in building simple theories. Although operational definitions may exist near the base of the hierarchy, soon we reach levels where all we require are rules of interpretation connecting this theory with theories on the next lower level. Thus the statements near the top connect with experience only through a long chain of conventions and mathematical arguments.

SUGGESTED READING

Complete references will be found in the Bibliography at the end of the book.

Operationalism.
  Bridgman, Chapter I.
  Feigl [1].

Concept formation.
  Hempel [1], pp. 39-50.

Rules of interpretation.
  Carnap [5].
  *Margenau, Chapters 4, 5, 6
  *Quine.
  *Carnap [3].

8

Measurement

"Are five nights warmer than one night, then?" Alice ventured to ask.

"Five times as warm, of course."

"But they should be five times as cold, *by the same rule—"*

"Just so!" cried the Red Queen. "Five times as warm, and *five times as cold—just as I'm five times as rich as you are,* and *five times as clever!"*

WE ARE FREQUENTLY TOLD that the fundamental activity in which a scientist engages is that of making measurements. Without measurements, we are told, no progress is possible in modern Science. We are also told that progress in Science is intimately connected with a refinement of our measuring instruments. In view of these far-reaching claims it is important to investigate the nature of measurements. This will be done in several steps. We will start with the very simple concept of classification and work our way up to a full measuring scale. To illustrate the procedure I will use two examples, one of which has already been worked out completely by scientists, that is, one in which we have a measuring scale available. This will be the example of temperatures, which we have already used as an example of explication. The other one will be an example that so far has not been worked through as far as measurement is concerned, but where we might at least see the direction in which developments must be made to give us a scale of measurement. This second example will concern the psychological attitude of a man toward Science.

CLASSIFICATION

There is no doubt that primitive man's approach to temperatures must be summed up by saying that he classified them roughly; perhaps as very hot, hot, moderately warm, lukewarm, cool, cold, and very cold. Naturally his classification may have been somewhat finer or rougher than this, but there is no doubt that he restricted himself to some sort of rough classification. Indeed, for most nonscientific purposes nothing more is needed than such a classification. Even these words are not necessary. One could simply describe temperature as being "like a day in May" or being "like a day in mid-July" or, on the other hand, being "like a day in the middle of the winter."

For such a classification to have any value at all, a few conditions must be satisfied. First of all, the classification must be such that it is possible to classify every single day into one of these several classes. The second condition is that no one day could conceivably be classified under two different headings; otherwise the resulting confusion would make the concept useless. The latter requirement is described by saying that the classes must be mutually exclusive, while the former is said to make the classes exhaustive. The advantage of such an exclusive and exhaustive classification is that we have reduced the problem of dealing with infinitely many days to the problem of dealing with a small number of classes. The days that have been put in the same class are considered as being alike as far as their temperature is concerned.

This last attitude is typical of classification. If we divide human beings according to their height, whether this be done roughly or to the nearest inch or the nearest hundredth of an inch, we have taken a very large number of human beings and put them into a few classes. These classes should be small enough in number to be manageable and yet sufficiently finely divided so that the human beings in the same class can be considered as having equal height.

In the last description we noted that the same concept can lead to many different classifications, finer or rougher. Just which one is most useful will depend entirely on the purpose for which we use the classification. For example, if the classification of human beings according to height is to be used simply to decide how large a bed the human being should purchase, a very rough classification into very tall, tall, short, and very short will be sufficient. If, on the other hand, one wishes to use the classification for the making of a suit of clothing, a finer classification is needed, and for police identification purposes it is desirable to have the measurement to the nearest inch or even some fraction of an inch.

The refinement of a given classification has a common procedure. We must be careful that the new classification will still be exclusive and exhaustive. To make sure that the new classification is exhaustive, we simply have to assure that all cases covered under the original classification are still covered. The easiest way of doing this is to take the old classes, one at a time, and subdivide them into exclusive subdivisions. This happens, for example, if we start with a classification of heights to the nearest inch and end up with a classification to the nearest quarter inch. Each of the old classes has been divided into four exclusive subclasses leading to a new exclusive and exhaustive, refined classification.

In many sciences we have not passed the stage of classification. For example, in many branches of medicine the doctor will be happy if he manages to classify a certain disease as belonging to one of a not too large number of classes. For each class he will have some information as to the correct method of treatment and hence classification is his major tool. For a simple question, such as whether to administer a sedative or not, a very rough classification may be sufficient. On the other hand, for a decision as to whether to use an antibiotic and the type and dosage of antibiotic, a much finer classification of diseases is needed.

Again, the student in an elementary psychology course will feel that the entire course is simply a course in "name giving." This is because much of psychology so far is restricted to a classification of personality types, intelligence groups, reactions to stimuli, etc. The student who reacts by saying that he has learned nothing of any use is nearly right but not entirely so. It is certainly the case that a mere classification will not have many practical applications; however, without a good classification most sciences would be unable to progress.

Let us now turn to our second illustrative example—the psychological attitude of a man toward Science. As far as classification is concerned, one might divide human beings into such very rough classes as: (1) likes all Science very much, (2) dislikes all Science, (3) likes theoretical Science but dislikes experimental Science, (4) likes experimental Science but dislikes theoretical Science, (5) likes both theoretical and experimental Science moderately, etc. The five classes given here are exclusive but are certainly not exhaustive. It would be a good exercise to try to complete these into an exhaustive classification.

Even as rough a classification as that just given has some useful applications. Suppose that a freshman adviser in college is able to put an advisee into one of these classes. He might then give him some useful advice as to what type of science course the boy should take, if any. Certainly a man who dislikes all Science should take a rather broad general education course, designed merely to teach him as much as any intelligent human being should know. On the other hand, someone who likes all Science very much should be advised to take a variety of sciences and perhaps be advised to work toward a major in the science division. Spotting men belonging to classifications three or four is one of the key tasks of a freshman adviser. It is very important for him to point out to such a student that it is entirely possible that he will be very fond of some of his science courses while he will dislike others. For example, a man in classification three would be well advised to elect Mathematics heavily and to work in such sciences as Physics and Chemistry, especially in some of their more mathematical branches. On the other hand, a man in classification four should be encouraged to elect the laboratory sciences and to keep his theoretical Physics and his Mathematics at the minimum required for his work.

To understand the full development of measurement and its significance in modern Science we must now see how a mere classification can be improved and extended.

PARTIAL ORDERS

The major task remaining to the scientist is trying to compare objects taken from different classes. For example, if we have divided days of the year into such different classes as "like a day in May" or "like a day in mid-December," we must try to compare a day that is like a day in May with a day that is like one in mid-December. In this case, since we are interested in temperature, we could take as our basis for comparison a statement as to whether one day is warmer than the other. This type of relationship must satisfy two basic requirements. First of all, if we say that April fifteenth is warmer than April fourteenth, we certainly want to exclude the possibility that April fourteenth is warmer than April fifteenth. In other words, if *a* is warmer than *b*, then *b* must not be warmer than *a*. In general, our relationship connecting up two classes will be of the form expressing that *a* is more so-and-so than *b*, or *a* is preferred in this respect to *b*. In either case, *b* must not hold the same relationship to *a*. This property is known as *asymmetry*.

So far we have looked only into the comparison of two objects. The next requirement arises when three or more objects are compared. Suppose that someone tells us the following facts: April this year was warmer than March; March was warmer than last November; and last November was warmer than April this year. This type of information would be useless to us. Indeed, if we are told that April is warmer than March and March warmer than last November, we want to be able to conclude that April is then warmer than last November. This property is known as *transitivity*. Similarly, in a question of preference, if *a* is preferred to *b* and *b* is preferred to *c*, we want to be able to conclude that *a* is then preferred to *c*. However, at this stage we allow the person to tell us that he is not sure which month was warmer or that he is not sure which object is preferable.

A relationship that is *asymmetric* and *transitive* is known as a *partial order*. It will allow us to arrange our various classes in an ordering where, although we cannot give complete information as to what class is ahead of which other class, some answers are supplied. One such set of classifications is illustrated in the figure.

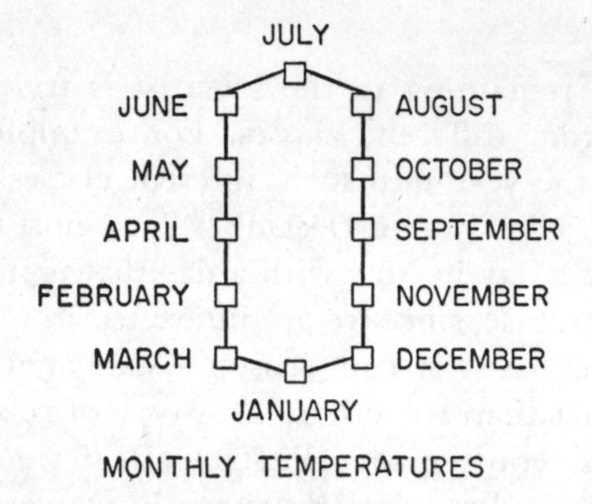

MONTHLY TEMPERATURES

ATTITUDES

As we see from this figure, we can definitely say that the month of July was the warmest month and the month of January was the coldest month. On the one hand, we can compare from memory the months from January to July and, on the other hand, the months from July to December and back to January, but we are unable at this preliminary stage to compare a month in the first half of the year with a month in the second half of the year. Nevertheless, such a simple classification can lead to generalizations. For example, in this case we can roughly state that the weather got warmer from January to July and then it gradually cooled off.

Let us return to our second illustration. In the classification of human beings according to their attitude toward Science, we may put at the top of our scale the class that likes all types of Science very much (class 1) and at the bottom of our scale the class that does not like any type of Science very much (class 2) while classes 3, 4, and 5 are on a level in between. A fuller classification, that is, an exhaustive classification, would lead to many more levels.

It is at this stage that scientists are likely to fall into serious traps. Although the property of *asymmetry* can be easily checked, the property of *transitivity* is more complicated, and there are many examples of relations that sound like partial order relations even though they do not turn out to be *transitive* in general.

For example, consider the competition of ten teams in a football league. We are tempted to order these, at least partially, according to the criterion of "Can *a* defeat *b*?" Unfortunately this is not a partial ordering relation, since often it is possible for the first team to defeat the second team, for the second team to defeat the third, and yet for the third team to upset the first team.

There are more subtle difficulties with partial ordering relations. For example, one would certainly think that making choices in a group according to majority decision is a good way of governing a body. However, this turns out not to be a partial ordering relationship in general. Suppose that there are three items among which the group is trying to choose. They would like to decide which one to purchase first, which second, and which third, according to the order of preferences of its members. In a poll of one hundred members, ten of them rate *a* first, *b* second, and *c* third. An additional twenty members come out with the order of preference of *b, a, c*. Twenty of them have the order *c, b, a*. Twelve of them have *b, c, a*. Thirty of them have *a, c, b;* and the remaining eight like *c* best, *a* second best, and *b* least of all. If we take these items two at a time we find that 52 members (a majority) prefer *b* to *a*. On the other hand, 58 members prefer *c* to *b*, and 60 members prefer *a* to *c*. Hence, we have a nontransitive situation in which it is impossible to reach a majority decision.

A particularly frequent example of a partial ordering relation is one in which the members are rated on more than one category. For example, suppose that we rate our employees both according to their age and according to their seniority. One man is to be rated ahead of another one only if he is both older and more senior. In this case we arrive at a partial ordering relationship. If *a* is ahead of *b*, then *b* cannot be ahead of *a*, hence we have *asymmetry*. Also, if *a* is both older and more senior than *b* and *b* bears this relation to *c*, then *a* must be older and more senior than *c*, hence we have *transitivity*. But it is not the case that we are able to compare all pairs. A man of age sixty who has worked for the company thirty years cannot be compared with a fifty-five year old man who has worked thirty-five years. This is a good example of a partial ordering relationship. Indeed, our second illustration is of this form. We have arrived at a branched-out tree in our diagram because we had two different criteria in mind—namely, liking theoretical Science and liking experimental Science—and we have made no attempt to compare these with each other.

In order to strengthen this concept we must attempt to rid ourselves of the undesirable feature of having pairs that cannot be compared.

SIMPLE ORDERS

The requirement that we would like to add to the previous ones is that, given any two objects from different classes, it should be possible to tell which one is ahead of the other. This feature eliminates the branching out of the diagrams that we have encountered so far. In a branching out we will always note that the difficulty is that two different objects on the same level in our diagram are not comparable. If they were comparable one would be ahead of the other one, and hence they would not appear on the same level. Since, according to our new requirement, no two classes can appear on the same level, our classes are now arranged in a straight line without any branch. This type of order is known as a *simple order*. Its three defining characteristics are (1) *asymmetry*, (2) *transitivity*, and (3) the fact that any two classes can be compared.

The temperature scale has this desirable attribute: it is possible to arrange days in the year from the very coldest to the very warmest without any case where comparison breaks down. This is achieved by means of our measuring instruments. Actually our measuring instruments accomplish more—namely, they assign a numerical value to the temperature of a given day. However, it is important to note that the scale—which is painted in a more or less arbitrary manner on our thermometer—is really not necessary to arrive at a simple order. If we simply noted for each day how high the mercury rose in the thermometer we would at the end of the year be able to arrange our days in order of how warm it was. It would not be necessary to say that on such-and-such a day the temperature was 52 degrees. It would merely be necessary to note that on this day temperature rose higher than on the previous day and hence this was a warmer day than the day before.

For many partial orders a simple ordering is easily achieved, since there is a definite trend, according to which the classes are arranged. A concept such as height or age of human beings is a good example, for we have a definite intuitive idea of going from youngest to oldest or of going from shortest to tallest. A less clear-cut example is that of colors. Here a classification into the basic colors of red, orange, yellow, green, blue, and violet is a natural procedure. However, the ordering of these is a much harder task. It took more sophisticated information as to the nature of light to lead us to a simple ordering. We had to arrive at the concept of the wave length or frequency of the various colors of light before arrangement in a simple linear order was possible.

In case the partial ordering was the result of two or more criteria being considered at the same time, we are up against a serious stumbling block trying to decide how these can be pulled together into a simple order. For example, in the case of age and seniority in a given company, the problem can be boiled down to asking, "What is the relationship between one year of age and one year of seniority?" For example, if we decide that one year of seniority is the equal of two years of age, then a man who is five years older but has five years less seniority than the second man will be placed behind him on our scale. Needless to say, many other formulas could be worked out (actually infinitely many) that would convert our partial ordering into a simple ordering.

How could we proceed with the problem of interest in Science? We could, for example, decide that, all things being equal, an interest in theoretical Science shows "a deeper interest" in Science than an interest in experimental Science. If this is our basic attitude we decide to arrange our five classes (see above) as follows: Class 1, class 3, class 5, class 4, and finally class 2.

It is the fact that we have so many possibilities available to us that is most puzzling, and here most of the difficulties arise. When a scientist makes a flat statement that it is impossible for him to arrive at a simple ordering, this generally means that there are too many possibilities.

NUMERICAL SCALES

The concept of measurement is closely related in our minds with that of numbers. A numerical scale consists of real numbers assigned to the objects under consideration. In most cases one can establish that a simple ordering preceded the assignment of numbers. The question now is just how to assign numbers to the elements in a simple ordering.

In the case of temperatures this is accomplished by means of the column of mercury in a thermometer. Here, someone noticed that as the temperature went up the length of the column expanded. Hence, the length of this column, which is a number, is used as a measurement for the temperature. In other words, we related our original phenomenon to another event where we had numbers available to us. This procedure is not as simple as it sounds.

First, in order to connect the length of the column of mercury with temperatures, we must have observed that there is a certain relationship between them. Specifically, we needed the law that the length of the column increases with the temperature and that there are no exceptions to this. We must note that for this very law we needed the simple ordering of temperature; otherwise this law could not have been expressed. In this case the resulting numerical scale would have been an impossibility without a previous simple ordering.

Secondly, we find many other objects that could have been used as measuring scales. Any metal will expand when exposed to heat, and the length of such a metal object could have been used as the measuring rod for temperatures. A temperature scale based on the expansion of an iron rod would be drastically different from the scale based on a column of mercury. The scientist constructing a temperature scale must have had some information indicating that the latter is better than the former. Specifically, there was a belief that the expansion of mercury is related in a much simpler manner to temperatures than is the expansion of iron. But just exactly what this statement means is not clear at the moment.

If we had an intuitive numerical scale for temperatures, then we could ask the question of how complicated the relation is between the length of an object and this given temperature scale. But if the length of this object is to serve as a definition of the temperature scale, any assertion as to simplicity is bound to be a circular assertion. However, the criterion for the choosing of such a scale can be resurrected in a new form. If we use the length of a given object as our measure of temperature, then the expansion of this particular object will be related very simply to temperatures. But we can then ask what we can say about the behavior of all other physical objects. The form that the law of expansion and other laws of heat will take depends on the kind of temperature scale that we choose. It is entirely conceivable that the laws look much simpler in terms of one scale than they do in terms of another scale. It is presumably this that led us to the choice of the currently accepted scale or scales of temperatures. This still leaves us with considerable choice as to the actual scales. For example, we have several different scales in use now, but these differ only as to the unit chosen and the place where we have assigned temperature 0. Such changes make minor differences in the statements of laws, but they do not change these laws drastically. In general, if one scale leads to simple laws, then any other scale of measurement related to the first one in a simple manner will lead to simple laws. We will say that these scales are not essentially different.

We may summarize this discussion by saying that the temperature scale arose out of a previous simple ordering, that there were many different possible choices, and the choice was made on the basis of simplicity and fruitfulness. Even when this choice is made there is still an arbitrary decision that differentiates between many different scales that are essentially the same. This procedure is typical of the formation of numerical scales.

Let us consider a different numerical scale, namely the scale for weights. The very word "scale" is intimately connected with weighing. The first step in the weighing process is the use of a balance for checking that two objects have the same weight. If an object A is placed in the left pan and B in the right pan, we determine their relative weights by observing whether the two pans stay at the same level. If they do, we say that the objects are equally heavy; if the left side is pulled down, then A is heavier, and a heavier B will pull the right pan down. This furnishes us with a simple ordering.

We are tempted to take this simple ordering as "obvious." But the fact that these relations determine a simple ordering is known as the result of thousands of years of experience. There is nothing *a priori* obvious about it. It could happen that A pulls B down, and B pulls C down, but when A and C are compared, it is C that pulls A down. It is not even obvious that if A and B balance and B and C balance, then A and C will also balance when compared.

The next important observation is that two like objects always weigh more (according to the above definition) than one such object. This suggests choosing some standard object for reference, and comparing a given object with several standard ones. If we assign weight 1 to the standard object, and if the object being weighed can outweigh 5 standard objects but not 6, then we assign to it a weight between 5 and 6. For many purposes such integer scales suffice. In a grocery story measurement to the nearest ounce will be entirely satisfactory.

To obtain a finer scale we may argue as follows: Suppose that three objects, exactly like A, together balance the standard object. Then it is reasonable to assign weight $1/3$ to A. In this way we may find objects that are small fractions of the standard object and use these as standard objects in a finer scale. Thus we find ounce objects, 16 of which balance a pound, and then we can measure to the nearest ounce rather than just to pounds. In other words, we can then express weights to the nearest $1/16$. But this method will give us only rational (fractional) number weights, and rational numbers are of a smaller order of infinity than all the real numbers. It is also doubtful whether we could even in principle make arbitrarily small weights, and hence, presumably, there will always remain a certain small error in our measurements. But that is certainly true even of measurements where we think that we can measure arbitrarily closely.

It is important to agree, however, that even if we know exactly what limitations apply to our actual weighings, we should not place any restrictions on the weights that may appear in our theories. For example, if we restricted ourselves to rational weights in our theories, we could not make use of the Calculus.

We must continue to emphasize the strong assumptions hidden in such a procedure. For example, if two objects like *A* balance *B*, why is it that we are *sure* that four objects like *A* will balance two objects like *B*? The fact that such assumptions have generally turned out to be valid have made progress in Physics much easier.

We might raise the question of what happens if we have a simple ordering but no means of introducing numbers. However, it can be shown that any simple ordering of objects taken from the real world can always be changed into a numerical scale. The difficulty is that this can be done in infinitely many ways. Hence, the real problem is to get some feeling for the fruitfulness of such a scale before making an assignment.

As an example let us return to the question of the psychological attitude of a person toward Science. Previously we have arrived at the simple ordering taking as the basic criterion the attitude of the man toward theoretical Science with experimental Science as a secondary factor. We could now easily assign numerical values, such as: If he likes all Science, four points; if he likes theoretical Science, three points; if he likes theoretical and experimental Science but only moderately, two points; if he likes experimental Science only, one point; if he dislikes Science, zero points. However, it is hard to see that such a scale would be at all fruitful. It might be more interesting if we could somehow measure the degree to which he likes Science, with theoretical Science giving a heavier weight than experimental Science. Such scales are actually constructed on the basis of interest tests commonly in use by guidance personnel. These numbers would be at least detailed enough to allow fine distinctions to be made and, on the other hand, would be chosen so as to give some promise of fruitfulness. For example, the number may indicate how the student compares with a future scientist at this stage of development. Perhaps an even more interesting number for a college guidance person would be an index that indicated how well a student with these interests is likely to do in various science courses.

It is often voiced that the factors we are here dealing with cannot be measured in numerical terms. We can now offer some justification to this criticism, for which there is some basis. Actually, many of these measures are completely arbitrary and give no promise of fruitfulness. The critics are at least right in saying that we have yet to find a really simple and fruitful measure of these various factors. In the case of temperature our justification was in terms of the simplicity of the laws of heat which are based on these scales. We cannot expect a final justification of various psychological indices until someone succeeds in basing simple laws on them.

We must not leave this topic without pointing out that for many purposes the simple ordering is just as useful as a numerical scale. While this is not true in the formation of theories, for the description of experimental results and for the classification of objects a simple ordering will normally perform all roles played by a numerical measure. Indeed, in many applications scientists distrust the procedure because they realize the arbitrariness of the scale used. In these cases restricting ourselves to a simple ordering would have been much sounder.

If for many purposes a simple ordering is just as useful as a numerical scale, we may ask why it is that scientists have such a strong prejudice in favor of numerical scales. The answer comes in the formulation of theories. Let us recall that theories are useless unless we can deduce interesting consequences from them. This deduction is essentially a mathematical process, and hence the form of the theory determines the type of mathematics to be used. A numerical theory will enable us to use powerful mathematical methods usually taken from the Calculus. A theory that is stated in terms of a simple ordering requires a much more intricate mathematical treatment. Indeed, the kind of mathematics needed for such a theory has been developed only in recent years, and many scientists are not familiar with it. Hence we find that the real reason for preferring a numerical scale is mathematical convenience. We may also foresee the day when, with the development of Mathematics, numerical scales will become much less important than they are now considered.

We thus see that, at present, measurement plays a central role in Science, because by assigning numerical values to various objects or phenomena in nature we are able to incorporate these phenomena in numerical laws. This in turn means that the full machinery of classical Mathematics is applicable. We also find that the emphasis placed on measurement is probably a temporary phenomenon subject to change as Mathematics develops.

SUGGESTED READING

Complete references will be found in the Bibliography at the end of the book.

Hempel [1], pp. 62-74.
Campbell, Chapter VI.
Cohen and Nagel, Chapter XV.
Russell, Part Four, Chapters VI-VII.
*Arley and Buck, Chapter 11.

9

Scientific Explanations

"Please would you tell me," said Alice, a little timidly, "why your cat grins like that?"

"It's a Cheshire-Cat," said the Duchess, "and that's why."

THE BASIC PURPOSE of Science is to form theories which will explain the facts of our universe. In the previous chapters we had to refer repeatedly to such scientific explanations and the time has come to consider the nature of these.

GOOD EXPLANATIONS

Let us study an example of scientific explanation. Suppose we ask a scientist why a book falls if we drop it. He would tell us that the book is an object heavier than air, and that all objects heavier than air fall if not supported. He would then be satisfied that he had explained the event.

The key to the explanation is that "all objects heavier than air fall if not supported." This is a general theory which has been well established by past experience, and hence is an accepted theory. But the theory says nothing about books. We must establish the fact that the book is an object heavier than air, before the theory applies. So we have to know that the book was an object heavier than air. Once we have the theory, and this fact, it *follows* that the book falls when dropped. Let us list these various factors:

(1) We must have general theories.
(2) These theories must be well established.
(3) We must be in possession of facts which are known independently of the facts to be explained.
(4) The fact to be explained must be a logical consequence of the general theories and of the known facts.

We started with a very simple example, but the pattern of an explanation is the same for the most complex case. If you ask a scientist how he explains the tremendous amount of energy released by the A-bomb on Hiroshima, his explanation may take several hours, but it will have the same four components. He will cite a large body of physical theories, dealing with atomic phenomena, the materials used, the engineering processes, and all the other laws that were used in designing the bomb. These are accepted laws, because past experience has led us to assign high credibility to them. In addition, he will supply us with certain facts as to the dimensions of the bomb, the amount of the various materials that went into it, and how these materials are distributed. Finally, he will prove to us (by means of a calculation taking several hours, using high-speed computers) that it follows from the theories that a bomb of the given specifications releases a tremendous amount of energy. If you look over this account carefully, and compare it with the dropping of a book, you will see that the same four features characterize both explanations.

Let us now consider an example from Biology to show that the type of scientific explanation just discussed is not peculiar to Physics. A biologist performs a breeding experiment on a species of plant and ends up with 312 tall plants and 88 short ones. His problem is to explain the outcome. He knows the characteristics of the plants with which he has started. The relevant characteristics are that he started with a number of tall plants and a number of short plants which bred true in the past (that is, the tall plants always had tall offspring and the short plants had short offspring). He crossed some tall plants with short ones, resulting in hybrids in the first generation. The plants we are now discussing were second-generation offspring. These are his known facts. The laws that he uses in explaining the outcome are Mendel's laws. These laws are well established, because many generations of breeding experiments have borne them out. One basic difference between Mendel's laws and the Law of Gravitation discussed above is that the former are of a statistical nature. This means that the theories involve probabilities and lead to predictions of the form "this event has such-and-such probability," rather than yielding categorical predictions. From the given facts together with Mendel's laws one can deduce that the biologist should have gotten 300 tall plants and 100 short ones. One can also deduce from these laws by standard probabilistic considerations that the actual outcome is unlikely to be exactly the predicted one, but is very likely to be near it. For example, a deviation of more than 30 in either direction would be most unlikely. In this sense we can say that the actual outcome can be deduced from the given theories and known facts. Hence, this argument qualifies as a scientific explanation according to our criteria.

The last example is by no means exceptional; it could even be regarded as the most general case. We have pointed out before that all scientific facts are subject to small errors, governed by statistical laws. Hence the best we can ever hope to achieve is to explain why something very much like what happened did happen. The same explanation would apply if a slightly different event took place.

The best way to understand a rule is to see it violated. Grammatical rules would be meaningless abstractions unless we experienced violations of these rules. Such violations, of course, are very common. Following this path, we will clarify our four rules by giving examples of violations for each of them.

BAD EXPLANATIONS

Let us suppose that we are trying to explain the fact that Joe is a man. We may offer the known fact that Joe is present in the room, and the general theory that all people present in the room are men. From this it will follow that Joe, himself, is a man. However, this is a bad explanation. It seems to be of the correct form, but actually Rule 1 is violated. The general theory that we offered is in fact not a general theory. It is a shorthand report of a number of isolated facts that we observed, namely observations on the sex of the small number of people present in the room. It is stated in a form that makes it appear as if it were a generalization, but it is not. In effect, we have first observed the sex of each person in the room, including Joe, and then we used this fact to explain that Joe is a man. Naturally, such an explanation is circular.

The same kind of incorrect explanation can be made more plausible if we use an explanation as follows: "Joe is a student at Dartmouth College. All students at Dartmouth College are men. Therefore, Joe is a man." When one generalizes about all students at Dartmouth College it certainly gives the appearance of stating a general theory. Indeed, if this is a statement about all Dartmouth students past, present, and future, then it *is* a general theory. If, however, it is merely a statement about students now present at Dartmouth College, or about students past and present at Dartmouth College, then it is a summary about individually observed facts, one of which is that Joe is a man. In the latter case, the explanation is again circular.

It is interesting to investigate the alternate interpretation of our theory, where the generalization is supposed to apply to all Dartmouth students, including future ones. In this case Rule 1 is certainly satisfied. However, Rule 2 is in considerable doubt. It is questionable whether any scientist is willing to state that past evidence is a good basis for accepting the generalization that Dartmouth will never admit women. We have seen too many incidents of strongholds of the American Man giving in to the pressure of the weaker sex. Hence, our explanation so interpreted would violate Rule 2. We now have a general theory but it is not well established. This is an interesting example where we have a bad explanation no matter how we interpret it, but one interpretation makes the explanation bad because it violates Rule 2, while according to an alternate interpretation it violates Rule 1.

Another example of the violation of Rule 2, and a clearer one, is: "All swans are white. Penelope is a swan. Hence, Penelope is white." In this case the general law is "All swans are white." The known fact is that Penelope is a swan. Let us accept as a matter of fact that Penelope is a swan. We cannot deny that the conclusion "Penelope is white" follows from the known facts together with the general law. But we now have conclusive evidence that the general law is false. Black swans were discovered in Australia and they are now to be viewed in many parts of the United States. Hence, our explanation is bad because our law is known to be false. This is an interesting example because the very same explanation was a good and generally acceptable explanation until black swans were discovered. This illustrates the fact that what may be an excellent scientific explanation in one century may be unacceptable in the next one. Just as theories have to be changed from time to time, or abandoned completely, so scientific explanations must also be altered or forsaken.

So-called scientific explanations provided by the man on the street usually violate Rule 2. If anyone has recently given you a complete explanation of the international situation, it is very likely that he made use of a generalization that he firmly believes, but which—in fact—has a very low degree of credibility. It is a common experience that laymen drawing conclusions from commonly known facts will arrive at entirely different predictions and explanations of events. This is due to the fact that each of them makes use of equally plausible and equally badly established theories. Under the same heading we ought to mention spur-of-the-moment explanations, such as, "Why does Jack dislike Spam?" "Because he had too much of it in the Army." The generalization which is hidden in this explanation, that all men who had a great deal of Spam in the United States Army will come to dislike it, is certainly not supported by the available evidence.

Next we will consider violations of Rule 3. When the fact is not independently known, then the explanation is circular. Suppose someone tells you, "Gloria is a tall blonde, and all blondes are attractive to men," and uses this as an explanation of the fact that Gloria is tall; you will certainly see that there is a circularity. But there may be more subtle examples. "All eagles have wings, and this is an eagle; hence it has wings." The trouble here is that it is part of the definition of an eagle that it have wings. Hence our original fact, that this is an eagle, contains the fact to be explained.

An important special case of this pseudo-explanation is the "name-giving" explanation. Suppose that we find an animal with the curious habit of walking in circles all the time, and we have no idea how to explain this. So we invent a new word, say "rotopedist," to apply to animals walking in circles. The next day we offer the explanation that the animal walks in circles because it is a rotopedist. Of course this violates the third condition. The fact used is "this animal is a rotopedist," which is the same as "this animal walks in circles," which was to be explained; hence the animal's path is not the only thing that is circular. But didn't we make use of the general theory "All rotopedists walk in circles?" We did make use of this so-called theory, however—walking in circles is part of the meaning of "rotopedist"—so the theory is actually empty (analytic). Thus our explanation is really circular.

Molière gives a delightful example of the name-giving explanation. In his play, "The Physician in Spite of Himself," his doctor gets a prize for explaining why morphine puts one to sleep. He points out that all soporifics put you to sleep and morphine is a soporific; hence morphine puts you to sleep. The first statement in the explanation is your general theory; the second is a known fact; and hence we claim that this is a scientific explanation. Of course, "soporific" is synonymous with "a drug that puts you to sleep," and thus we again have an empty theory and a circular explanation. But the doctor may rest assured that he was not the only one in the history of Science who became famous for thinking up a new name.

Newly developed sciences are particularly guilty of substituting name-giving for explanations. Students electing an introductory course in some of these fields are frustrated by this phenomenon. One must, however, give the sciences some credit for their procedure. We have noted before that it is necessary to do a large amount of name-giving in a new science, since classification is a fundamental process in all of Science; and it is the beginning of the formation of scientific explanations. However, one must not confuse classification with explanation.

The final type of pseudo-explanation is the *non sequitur,* where the conclusion just does not follow. Suppose you point out that a certain lady is a rather ineffective Congresswoman. Someone volunteers the explanation: "Look, she is a woman, and all women have a weakness for men, that is why she is ineffective in Congress." It is true that she is a woman, and it is fairly credible that all women have the specified weakness, but absolutely nothing follows from this about her effectiveness as a legislator. Such a violation of Rule 4 is a favorite trick of political orators. When asked to explain an embarrassing event, they will quote impres sive figures and give a very plausible theory; you are not supposed to notice that these are entirely irrelevant to the event to be explained.

Sometimes, however, the irrelevance is only apparent, because some step of the explanation was omitted. We must now look at these partial explanations.

INCOMPLETE EXPLANATIONS

In a complete explanation the event to be explained is deduced from certain theories and from known facts. In an incomplete explanation some theory or fact is omitted. Let us first consider explanations using only a single theory and a single known fact; in this case the incomplete explanation either has no theory or no fact.

Suppose that we have an explanation with no fact in it. It might be of the type "All men are muddle-headed, that's why Joe is muddle-headed." The missing fact is that Joe is a man. This is rather typical of incomplete explanations. The missing factor is so obvious that we do not bother to supply it.

There is an analogous case where a rather obvious theory is omitted. For example, "Joe is human, hence he makes mistakes." The missing theory, of course, is that all humans make mistakes. But in the case of theories this justifiable omission is by no means the general case. Usually the explanation is lacking a theory either because we do not know what it is, or because we do not care to state it. Someone makes one of those so-called Freudian inferences: "Bill hated his governess, that's why he is shy." We have a vague feeling that this is a reasonable explanation and that Freud's theory is the missing link; but if we try to fill the gap, we run into considerable difficulty. The obvious bridge is "everyone who hates his governess grows up to be shy," but this is surely not true. If, instead, we try to make a weaker theory "everyone who hates his governess and finds no outlet for his hatred grows up to be shy," the theory may be slightly more credible, but it no longer serves the purpose. If we add it to the explanation, then we need more facts, namely, to show that Bill had no outlet for his hatred; we have just shifted from one kind of incompleteness to another.

A particularly interesting case of such incomplete explanations is one in which the author of the explanation would be shocked if he realized what theory he has taken for granted. For example, we will take the college student who says that he votes Republican because his parents vote Republican. What could the missing theory be? The most natural guess is that the missing theory is "I do everything that my parents do." This may, as a matter of fact, be the correct missing theory; however, the student would be the last person to admit it. The alternative to this distasteful theory would be something like "Everyone votes the same way that his parents do," which is certainly false. There may be other alternatives, but in each case we will find that the missing theory is either false or distasteful to the author of the explanation. This is a good example to point out why it is so important to make our scientific explanations complete whenever this is feasible.

There are many more complex cases of incomplete explanations. As a matter of fact, if you asked a scientist to explain a certain fact, you are most likely to get only an incomplete explanation. The theories of Science are so strongly interrelated that even a relatively simple explanation may make use of all the theories of at least one given branch. When a biologist explains a certain fact about microorganisms, he is likely to use the observations made through a microscope and his biological theories, but he is unlikely to state the optical laws governing the microscope, even though these are needed for a full explanation. This is why it is sometimes hard to judge whether a scientific explanation is adequate; we are to assume that all the missing steps can be supplied in a perfectly straightforward manner. But if the explanation is challenged, there is no substitute for actually bringing it into a form satisfying the four rules.

EXPLANATIONS VS. PREDICTIONS

We must now raise the question of whether our explanations really do explain facts. For example, we have said that the falling of a book is explained by the fact that it is heavier than air and the theory that objects heavier than air fall when not supported. Does this really explain anything?

It is true that the explanation satisfies all four of our conditions. But why do all objects heavier than air fall? We could explain this theory in turn by referring to the Law of Gravitation. But why does the Law of Gravitation hold? Clearly there is no end to these questions. This phenomenon is a common one in experience with children. No matter what answer we give them they will come back with a new "Why." For three or four stages the parents will be patient and will answer the children's questions, but soon the children will exhaust a parent's knowledge of theories and the parent will then have to come back with "Children should be seen and not heard." If the child persists in coming back with an additional why, the time has come for the parent to demonstrate his superior physical strength.

In this sense no complete explanation is ever possible. According to our four rules we must always make use of an accepted theory, and we may always ask the question why this accepted theory holds, leading to an infinite regress. Nevertheless, scientific explanations accomplish a useful purpose. Our explanation amounts to showing that the new fact fits into the general pattern of knowledge available to us. It is, in a sense, just what we would have expected. To this understanding of an explanation it is irrelevant why we would have expected just this. It is sufficient to know that we could have expected something of this sort to happen. When we put the matter in this form, we are coming very close to talking about predictions. When we say that we could have expected this event to take place, we mean that we could have predicted it. On the basis of past experience we formulate a general theory as applying not only to the past and the present, but also to the future; and on this basis we made, or could have made, a prediction to the effect that the event in question would take place. This points out a strong similarity between predictions and explanations, which is worth pursuing further.

Let us ask what the essential difference is between an explanation and a prediction. There is a rather spectacular difference in that an explanation refers to something already known to be true, while a prediction is a commitment to knowing what is going to happen in the future. This may be a difference striking from the outside, but if we look at the internal machinery of explanations and predictions, we fail to find any essential difference. In either case, we have a general theory available which must be well confirmed. We have facts which we can start with, and from the theories and facts known to us, we deduce a certain new fact. Here the word "new" may mean "new to us," or "not yet taken place." As far as the logic of the problem is concerned, there is no essential difference.

Many people were amused to read that one of the famous, high-speed computers has succeeded in predicting a storm that took place several years previously. Actually, this was a very major achievement. In making the prediction, the scientists used only known theories from meteorology, and facts that were known *before* the storm. During the year that the storm actually took place, both these theories and the facts were known but, due to inadequate means of drawing logical conclusions (that is, computing machines), the meteorologists did not predict the storm in time. The computing machine made no use of "unfair" information that was not available to the meteorologist ahead of the occurrence of the storm. The fact that the actual computation was carried out years later is entirely irrelevant to the procedure. This example shows that whether a given logical deduction is an explanation or a prediction depends on the purely accidental feature as to whether it is carried out before the event occurs or afterward.

The logical identity of the process of prediction with the process of explanation serves as a useful check on whether an explanation is scientifically acceptable. We can always ask ourselves the following question: "If we knew these theories in time, and had all presently known facts available to us, could we have predicted that this particular event, or something very much like it, would take place?" If the answer is "Yes," then we have a scientific explanation. If the answer is in the negative, then we are rationalizing.

This analysis throws new light on the various stages of the scientific method discussed in Chapter 5. At the third stage, the scientist must deduce certain predictions from his theories and known facts, and in the fourth stage he must verify these. The implication in this description is that these facts are somehow entirely new. It is now clear to us that it is just as valuable for a new theory to be able to predict a previously known but unexplained fact as to predict something that is yet to take place. A famous example of this is one of the major pieces of evidence provided for the Theory of Relativity. Newton's Gravitational Theory gave excellent explanation of the paths taken by the various planets, with one notable exception. His predictions for the path to be taken by the planet closest to the sun, Mercury, did not agree with future observations. Einstein's Gravitational Theory, on the other hand, was able to predict or explain exactly why Mercury's path did not agree with Newton's original predictions. This result was less spectacular than Einstein's prediction of the bending of rays of light, but it served just as well to confirm the Theory of Relativity.

There is a lingering feeling that somehow it is better to predict the future than to explain something already known. The reason for this is that it is so easy to offer a pseudo-explanation of a fact that we already know. However, the criterion here must be whether the four rules have been satisfied and not whether the fact was already known. The pseudo-explanations can be spotted either because they are circular, or because the theories they make use of are insufficiently confirmed, or because the argument is incomplete. The reverse situation is also frequent in Science. It may turn out that the scientist makes a spectacular prediction which is, as a matter of fact, verified in the future, and nevertheless it turns out that his explanation was unacceptable. In such a case he has substituted a fortunate guess for a scientific explanation.

THE HIERARCHY OF EXPLANATIONS

In order to be a good scientist, one must never lose the childish urge of asking "Why." Even if we know that an answer will only lead to more questions, the scientist never stops asking. Some of the most successful branches of science illustrate the progressive answers to repeated questions of "Why." Let us look at the laws of motion.

First of all there are certain laws which are derived "directly" from experience. Among these we might list the three laws of Kepler describing the motion of planets, Galileo's law of free fall, and the law of the tides. All these explained certain known facts. Then Newton asked "Why," and constructed a theory which explained all three of the previous theories! This process is similar to that governed by the four rules, only we use it to explain a theory, not a single fact. In other words, what we deduce is not an isolated fact, but a general theory. From Newton's various laws (and a few known facts) we can deduce all the laws of Kepler, Galileo's law, and the law of the tides. Later Einstein formulated his General Theory of Relativity, from which he could deduce not only Newton's laws but, for example, laws about the motion of rays of light. Thus we get a hierarchy of explanations, where the facts on the lowest level are explained by theories, and then each theory in turn is explained by the theories on a higher level, until we reach the limits of our present knowledge. (See the diagram.)

There are several important features that we must note in the hierarchy. It is not quite true that Kepler's laws follow from Newton's. Actually Newton derived an *improved* law of planetary motion! He showed that the planets would move in ellipses if there were no other planets, but that the sister planets pull them out of their elliptic orbits. Thus Newton could explain not only the fundamentally elliptic paths, but also the deviations (with the exception of a small error in Mercury). Similarly, he showed that

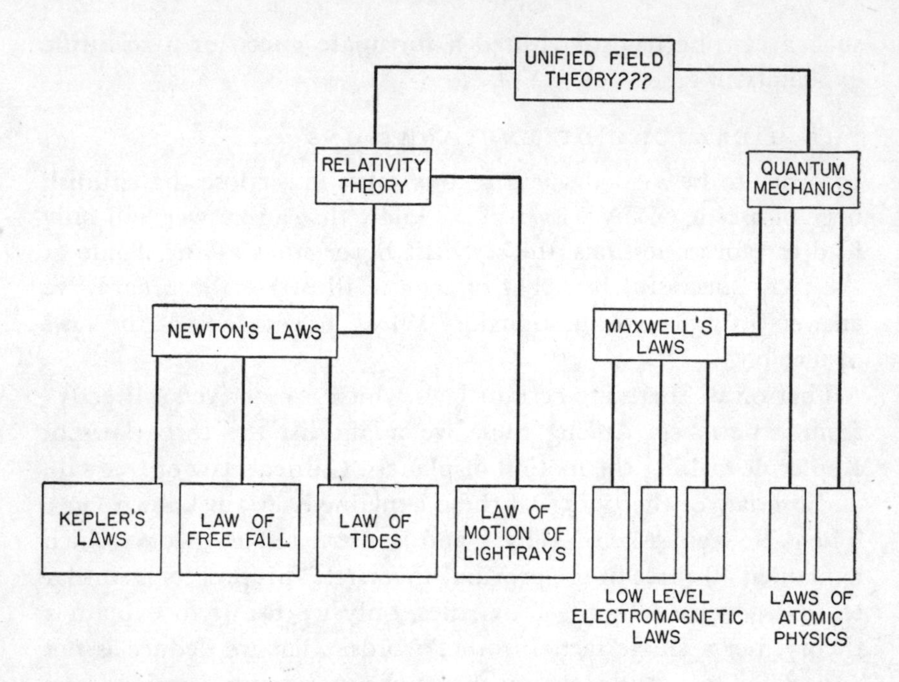

HIERARCHY OF LAWS

Galileo's law is only approximate, since the acceleration of a stone increases ever so slightly as it nears the surface of the earth. Quite analogously, Einstein could show that Newton was approximately correct, but he could also improve on him, for example, in the path of Mercury.

The second notable feature is that the "higher" theories cover more ground than all the lower ones put together. Newton could explain, in addition to the motion of planets, free fall, and tides, also the motion of comets, or of bullets. Einstein could predict the bending of light rays, undreamed of before. So the urge to explain leads to progress in two directions, greater accuracy, and wider applicability of the scientific theories.

We now have a picture of the progress of Science as being able to explain facts in a wider and wider field, and more and more accurately. We could thus picture the scientist as striving to the goal of finding The Law of Nature which would enable him to explain all facts with perfect accuracy. Even if this goal is unreachable, the scientist does strive to find a theory with applications in all branches of Science—a single theory which will explain all his other theories.

Even if a single law were found which unifies our body of knowledge, this need not be the end of progress. We would still strive to improve this law in both of the specified directions: broaden it, and make it more accurate. There are also likely to be unsolved mathematical problems at every stage of Science, so that the improvement of theories must be paralleled by the increasing mathematical development of existing theories.

This last point is illustrated in the development of Einstein's theories. For the Special Theory of Relativity all mathematical tools necessary were already available. For the General Theory most of the necessary mathematics had been discovered, but Einstein was actually not aware of this, and he had to rederive a good part of tensor analysis in order to be able to draw conclusions from his General Theory. In his new Unified Field Theory we are in the unfortunate position that the mathematical problems raised by it are much too difficult for present-day Mathematics; hence actually no conclusion can be drawn at all, no predictions can be made, and no test can be applied as to the power of the new theory to explain known facts. This is the reason why scientists have to take a neutral attitude toward the new theory. Here we have an example where further development of Physics must wait until Mathematics catches up with it.

THE ROLE OF "PURPOSE" IN EXPLANATIONS

In our minds explanations are related to the concept of "purpose." If we ask someone why he did a certain act he will tell you what purpose he had in mind. He went to the movies because he wanted to relax. He read a detective story because he hoped to discover an exciting plot. He went to a certain college because he hoped to acquire a certain type of education. In each case there was a desire, a hope, a certain purpose that he wanted to fulfill. This lingering connection may be at the root of our dissatisfaction with many scientific explanations.

In Greek Science the desire to explain everything as arising out of a purpose was so strong that they read "purpose" into inanimate nature. The resulting explanations were most picturesque, but scientifically worthless. Their explanations always amounted to saying that something took place because it was in agreement with the purpose of the universe. This sort of explanation would be useful if we had some idea as to what purpose the universe had. In so far as no precise information was supplied by the Greeks on this point, their explanations had no scientific value. The twentieth-century version of this type of explanation is to say that a certain event took place because God willed it. This statement may or may not be true; however, it has no value as a scientific explanation unless we know what God's will is. Earlier we pointed out that there is a simple test for a scientific explanation—namely, whether we could have predicted the event with the given theory and known facts. If we had known everything that led up to this particular event, and we knew that events take place only if God willed them, could we then have predicted the particular event? The answer is certainly "No." This is what we mean when we say that this type of religious explanation is of no scientific value, no matter how important these explanations may be in theology.

The interesting feature of both the Greek and the religious explanation is that they do not serve as scientific explanations unless we know what the purpose of the universe or the will of God is. And if we know this purpose or this will, we no longer have to make use of the given type of explanation. Suppose we had some partial knowledge of God's will, say that He wills that there should be a major war once every twenty-five years. This general theory would certainly serve for the purpose of scientific explanations or predictions. Suppose that twenty years have passed without a major war, we could then predict that there is going to be a major war within the next five years. We now have a correct scientific explanation, but the relevant part of the explanation is "There is a major war at least once every twenty-five years," and the fact "there has been no major war for the last twenty years." That the former is part of God's will played no role in the explanation.

We may now be criticized for claiming that human beings' purposes are irrelevant to Science. Here, unfortunately, we are caught in a difficult linguistic problem. The word "purpose" is used, on the one hand, as God's purpose or the universe's purpose, and as the purpose of a human being on the other hand, in what may be related senses, but certainly are not identical meanings. No one can deny that the purpose of a given human being, his desires, his hopes, his intentions, are highly relevant for Science. The role they play, however, is not that they are substituted for The Law of Nature, but that a human being's purpose is one of the facts that has to be taken into account in making scientific predictions.

We find a modification of the universal purpose type of explanation in teleological explanations. Here it is no longer claimed that every last detail of every event in the universe is determined by a universal will or purpose, but it is claimed that the general direction in which the universe tends is determined, and this fact is used to explain events. Here we find the type of explanation where the general theory states a certain goal toward which nature or some part of it tends. This sort of explanation is historically connected with the concept of purpose, but there is no logical necessity for this. In the third chapter we saw that actually our most useful laws possess a double direction. They can be looked at as being both causal and teleological. Given data about the present position of, say, a planet, and about its velocity, we can not only predict its future position but we can also deduce where it was in the past. This can be done by means of Newton's laws which are differential equations. Yet it would be hard to say that Newton's laws somehow embody a mysterious purpose.

Actually there is a version of many of the physical laws that is particularly suited to be described in teleological terms. This is the version in which laws are described by stating that nature behaves in such a way as to minimize or maximize a certain quantity. Knowing that the outcome of certain events will be to minimize a given quantity, we can predict at least within certain bounds just what the intermediate steps will be. Thus we will have a good example of a teleological explanation. Yet it is hard to identify this with a conscious purpose. Indeed, causal and teleological explanations are so closely interwoven that it is hard to find a pure example of either one. Perhaps Mendel's laws are as close as we can come to a purely causal explanation. They enable us to predict statistically the next generation in an experiment in breeding, but they do not furnish any useful information on the previous generation. Actually even this is not entirely accurate. It would be more accurate to say that, while Mendel's laws furnish a great deal of information about the future generation, they give us only partial information about the past. This is illustrated by court cases involving questions of paternity. It is much easier to predict the genetic make-up of a child on the basis of the genetic make-up of the parents than it is to go in the reverse direction.

For a purely teleological law we might perhaps select the law of Entropy. This law tells us that we will eventually reach a state of complete disorder and that nature tends in this direction, but it does not tell us anything very useful about the next moment or the moment just passed.

We are confronted with a certain phenomenon, and we cannot find the exact law. We may then glimpse in one of two directions: We may guess how it develops from one day to the next, without being able to guess the long-range outcome; or we may guess the eventual outcome, without understanding the day-by-day proceedings. Either (if true) is a good scientific theory, and hence legitimately usable in explanations.

Science should never be hampered by philosophical prejudices as to what type of explanations it should use. Any well-established theory is a great asset to Science and is of potential value in explanations. Just what form this theory takes depends on our ingenuity and our limitations. Since the latter are far greater than the former, let us not limit ourselves further by arbitrarily binding our hands in the formation of theories.

SUGGESTED READING

Complete references will be found in the Bibliography at the end of the book.

Explanation.
*Hempel and Oppenheim.
Frank [2], Chapter 6.
Campbell, Chapter V.
Duhem [2].
Schlick.

Hierarchy of explanations.
Feigl [2].

Teleological explanations.
Pap, Chapter 11.
*Nagel [3].

10

What Is Science?

"Are we nearly there?" Alice managed to pant out at last.

"Nearly there!" the Queen repeated, "Why we passed it ten minutes ago!"

ONE NATURAL WAY to start a book on the Philosophy of Science is with a definition of Science. However, such a definition would have to be superficial. A much better definition can be given after a great deal of other material has been clarified. By this time the reader is probably ready to ask Alice's question, "Are we nearly there?" We are now in a position to give the same answer that the Queen gave to Alice.

WHAT UNITES SCIENCE

Our alternatives are to define Science by its subject matter, or by its method. But the purpose of Science is to study the whole field of factual knowledge; it has no special topic of its own. Yet we certainly do not classify every study of facts as Science. For example, we refuse to admit Astrology into the family of sciences. Astrology studies facts; it studies the position of stars, and various events in human life, and tries to establish connections between them. The reason that we reject it as a science is not due to its subject matter, but because we consider the methods used by astrologers unscientific. Whenever we find a branch of supposed factual knowledge rejected by Science, it is always on the basis of its method.

Let us look at the other side of the argument. Is every application of the scientific method really a case of Science? The argument to be presented is that every such case can legitimately be called Science, though it is admitted that sometimes this is disputed. There are two types of applications of the method which many scientists would consider outside Science: every-day-life applications and applications by "nonscientists" like criminologists. Custom prevails in these cases: they are not sufficiently "dignified" to be classed as sciences. Perhaps it would be best to state the definition of "Science" to include only important applications of the method, but there is no good way of defining what an important application is. I find it best to consider Science in the broad sense, and call applications to every-day-life Science on an elementary level.

No one can *prove* what the right way of defining "Science" is, but we can argue about the most useful way. Since there is disagreement about the use of the word, our definition cannot agree with all different uses, but we can have several guiding principles: (1) Whenever there is a consensus as to whether some field belongs to Science, our definition must agree with the accepted verdict. (2) In cases where there is considerable disagreement, our definition must settle the dispute. (3) Of the many different ways of settling the disputes, ours must be one that leads to a useful concept. (4) The definition should be as simple as possible. These are the four conditions for the explication of a vague intuitive concept (see Chapter 6). I feel that defining Science by its method is the only definition satisfying these conditions.

The definition of Science by its method certainly agrees with usage whenever there is a clear-cut agreement; it is a simple definition, and it decides disputed cases according to an important principle (whether the procedure used was according to the rules of scientific method), and hence leads to a useful concept of Science. For this reason I shall use the word "Science" to be all knowledge collected by means of the scientific method.

The principal rival to the definition here given would be a definition according to common usage. In this we would poll the leading experts to tell us how they use the word "Science." Such a definition would suffer from all the prejudices of practicing scientists and would be sufficiently vague to destroy its basic fruitfulness. It is a useful precept in explication to err on the side of simplicity rather than on the side of common usage. For this reason I shall reject the definition by common usage and adopt the definition according to method.

There is an apparent circularity in this definition. Doesn't the definition of "Science" use the term "scientific (method)"? It does, but there is no circularity, since the definition of the method was given without using the term "Science." This is a very common procedure; for example, we are likely to define "Mathematics" as the study of mathematical laws, and then we give an independent definition of what such a law is; or we would define "happiness" as the feeling experienced by a happy person. We defined the scientific method by the cycle of induction, deduction, and verification, and by its eternal search for improvement of theories which are only tentatively held. Nowhere in this definition have we used the term "Science," so we can use the definition in turn to define "Science."

Just as a check on the definition we might inquire whether this method is really used in all branches of Science, and the best way of seeing this is through examples chosen from a variety of branches. The example discussed at length in Chapter 5 (discovery of Neptune) was from Physics. From Chemistry we might select the history of the Phlogiston theory. It was held for some time that burning consists in the giving off of a substance, and this substance was called phlogiston. But if this theory is right, then burning should reduce the weight of the burning material. When Lavoisier showed that, on the contrary, it gains in weight, then the old theory was rejected and the new theory (that a substance is taken in during burning) was formed to explain the facts. From this theory many consequences were deduced, and verified. For example, in a closed space there can be only a limited amount of this substance (oxygen) in the air, so when this is used up nothing more can burn in this space until fresh air is let in. We can easily verify this by putting a candle under an inverted glass, and watching it go out long before the candle is completely burned. This completes one cycle of the scientific method.

From Biology we will choose the discovery of Mendel's laws. Mendel observed over a period of years the variety of plants that he grew in his small garden. From these he formed generalizations concerning the proportion of various traits in the offspring, as determined by the traits of the parents. From these laws we can make sweeping predictions as to the results of breeding experiments, which have been repeatedly confirmed and have led to considerable profit for farmers throughout the world.

It is harder to find good examples of the scientific method in Psychology. However, recent developments in Learning Theory will furnish us with examples. Very interesting theories have been developed by R. R. Bush and F. Mosteller, on the one hand, and by W. K. Estes, on the other hand. These theories start with data collected in simple experiments in which the experimentor attempts to teach a rat, a goldfish, or a human being to learn how to do a task. The two theories provide alternate (and related) models as to how the subject learns to perform these tasks, and from these theories predictions can be made about the outcome of the experiments. Predictions may concern the average time it takes a subject to learn the experiment or the number of errors he will commit before learning how to do the task perfectly, or even whether he will ever learn to do the task perfectly. In the simplest cases these predictions are in excellent agreement with experiments.

The movies provided us with a good case from the Social Sciences. The record of the Kon-Tiki expedition is a gallant testimony to the future the scientific method has in this field. Certain similarities between the ancient traditions of natives in the South Sea islands and of inhabitants of South America led a group of sociologists to form the theory that these natives came from South America, not from the much nearer shores of Asia. This theory was disputed, because it seemed impossible that a thousand years ago these primitive people would have been able to undertake such a journey of several months over the open sea. The scientists deduced from their theories that the type of primitive craft (a loosely constructed raft with poor facilities for steering and only a limited place for storing food) must be capable of completing such a journey. They risked their lives to test this theory, by actually attempting the trip on just such a raft. They found out a great deal not known before. They found a plentiful supply of palatable fish in these unknown waters, they found that the looseness of the raft prevented flooding, and they found that, while steering was impossible, the prevailing currents carried them unerringly to their destination. Over a hundred days later they arrived on these islands and were greeted by the natives who related their old legend according to which the great god Kon-Tiki had brought their ancestors to the islands on just such a craft. Thus they verified their theory (which previously was in general disrepute) and completed a thrilling chapter from the history of the application of the scientific method.

WHAT DIVIDES SCIENCE

We have seen that Science is united not by its subject matter, but by its method. We will now see that it is the subject matter that divides Science into branches.

It is very difficult to explain the reasons for dividing Science the way we do, especially since there is no really good reason. Let us compare the various branches of Science to the various colors. There are five basic colors—red, yellow, green, blue, and violet. Someone else will tell you that there are six or seven, adding orange and/or indigo. That still leaves us with unclassified mixed colors, like pink or brown, not to mention white which is a mixture of all other colors. Even among the basic colors it is difficult to classify all shades. Let us pick a definite shade of blue-green, and we will get quite an argument as to whether it is green or blue. It is even more difficult to distinguish between shades of blue, indigo, and violet.

Quite similarly, we will get an argument as to what the fundamental branches of Science are. Physics, Chemistry, Biology, Psychology, and the Social Sciences form a common list, but others will add Astronomy, and divide the Social Sciences into Economics, Sociology, and Politics. That still leaves us with "mixtures" like Biophysics or like History (insofar as History is made scientific). We have borderline cases too; for example, it is sometimes difficult to say of viruses whether they are alive (and hence belong to the subject-matter of Biology) or whether they are inanimate molecules (and hence belong to Chemistry).

In the case of colors we know that there is a continuous scale, and that explains why there is no natural division into five, seven, or any small number of distinct colors. In addition, we can get an endless variety of secondary shades by mixing the primary ones. We know that it is the difference in wave lengths that differentiates between them, and hence it makes sense to speak of one (pure) shade being closer to a given shade than to another. But any division into colors is highly arbitrary. We do not have as complete a picture of the structure of Science, since our knowledge of theories is incomplete. Scientists will study a group of phenomena which seem related, and try to connect them by means of a theory. Sometimes they fail, and at other times they succeed. In the latter case we have a branch of Science. But there is a great deal of arbitrariness in this procedure. After all, we know that *all* phenomena are connected through The Law of Nature. The laws we are looking for are partial laws, and we are likely to find them where we look for them. Kepler would never have found a connection between the falling of an apple and the motion of the planet Mars, because he never looked for such a connection. It was left to Newton to connect Astronomy with Physics. The ancient Greeks saw a sharp boundary between them; in the heavens circular motion was "the law," while on earth things moved in a straight line. As we learned more, Astronomy became more and more incorporated into Physics. There are still unexplained phenomena, so there is still some excuse for an autonomous Astronomy.

Whenever there is a great deal of arbitrary choice, accidents take a hand in the decision. Whether your university has a separate Astronomy department may depend on whether some rich alumnus has a secret passion for Astrology, but, not being allowed to endow a chair in fortune-telling, he leaves a few millions to finance an Astronomy department. Or, on the other hand, the professor of Astronomy may have the ambition of becoming head of the entire Physics department, and hence there is a merger. I even know of a strange case where Astronomy is part of the Mathematics department.

We witness similar factors in the ordering of shades of color. Suppose you like blue, but dislike green as a rule. If confronted by a shade of green-blue, you are likely to classify it blue or green, according to whether you like the shade or not. Historical factors may have a considerable influence too. If the same person works on two apparently different types of phenomena, these may both be assigned to the same science. Perhaps the fact that Newton studied both Mechanics and Light may account for their becoming branches of Physics, while the fact that he did not contribute to the study of chemical compounds may account for the independent status of Chemistry.

We may summarize the foregoing discussion by stating that Science is divided into branches arbitrarily. If phenomena are connected by known laws, or if some scientist attracts sufficient interest in the study of these phenomena, or for a number of accidental reasons, a group of phenomena is collected into a branch of science. It is dangerous to place too much emphasis on such arbitrary divisions.

REASONS FOR DIVIDING SCIENCE

There are two schools of thought according to which one can motivate the division of Science into branches.

One school believes that eventually a unified Science will be possible and has definite views as to how it will come about. This school will give a division of Science somewhat as follows: Physics, Chemistry, Biology, Psychology, the Social Sciences. They feel that through the progress of Science the "higher" disciplines will become branches of the lower ones and eventually all of Science will be *reduced* to Physics. The Social Sciences are to be reduced to Psychology by explaining the actions of a group on the basis of the individual psychology of its members. Psychology is to become a part of Biology and the workings of the human mind explained in terms of the workings of the human body. The human body again is to be considered as made up of certain chemicals and subject to various laws of Chemistry, and hence Biology is to become a branch of Chemistry. Finally Chemistry is to be reduced to Physics, a process that is already fairly well completed. In this manner all of Science will be united as a great, expanded Physics.

It is pointed out that in each case the behavior of the whole is explained in terms of the behavior of its parts. This fact is considered significant by this school of thought. Human groups are to be considered as wholes made up of individual parts. These individuals are in turn made up of cells, the cells are made up of chemicals, and these in turn are made of atoms. Many precedents are cited where the behavior of wholes has been explained by Science in terms of the behavior of their parts. It is further pointed out that the lower sciences have wider applicability than the higher ones. For example, the theories of Physics have universal applicability and are used by all sciences. The Law of Gravitation applies equally well to a stone or to a cat, when one of these objects is thrown out of a window. However, Mendel's laws are applicable only to the cat and not to the stone. Thus the ordering given above leads from more general theories to more specialized ones.

We now have a threefold reason for dividing Science, and for ordering the branches in the given manner. First of all there is the expectation of reducing the later branches to the earlier ones. Secondly, the later branches treat wholes whose parts are treated by earlier branches. Finally, there is successively more specialization as we go through the order of the branches.

Although this kind of division certainly is reasonable for many purposes, it must be pointed out that the reasons given are highly oversimplified. For example, the manner in which Biology makes use of the Laws of Gravitation is entirely different from the manner in which the Social Sciences make use of Psychology. The Law of Gravitation will apply to a biological object as a whole, as well as to any one of its parts. However, psychological laws will apply only to the parts of a social group, certainly not to the whole, unless we decide to use "Psychology" in two different senses. Again there is no good reason why the lower sciences should not utilize theories from the higher disciplines. For example, while the Theory of Evolution makes use of Geology (which we may take to be a mixture of Physics and Chemistry), Geology in turn is helped out by results taken from the Theory of Evolution. One of the most useful tools in modern Geology is the dating of rocks by means of the fossils contained in them. The ordering presented is very neat as long as we do not look at too many examples that fail to fit into this order. For example, we might ask whether Nuclear Physics is a separate branch, and we would have a very difficult time placing Astronomy into the neat order. Even the basic principle that wholes are reduced to their parts must be taken as a great oversimplification. In what sense are the objects of Economics reduced to a study of their parts, and how can the reduction of any objects of the higher disciplines to a field theory (such as the Theory of Relativity) fit into this scheme?

Oddly enough, a second school of thought supports a division of Science precisely because they do *not* believe in the possibility of reduction. Vitalists will separate Biology from the lower sciences because they believe that, in some sense, there is a sharp boundary between them. This will be discussed in Chapter 12. Similarly, dualistic philosophers will maintain a division between Psychology and Biology as we will point out in Chapter 13.

I do not want to enter these disputes at the present moment. Let us be satisfied with stating that, for many purposes, a simple division of Science into branches is very useful, but we have found no sufficient reason for assigning deep significance to this classification. We conclude that Science is an enormous area of human research which is united by a common method. Its divisions are for convenience in describing results and do not represent a fundamental feature of Science.

SUGGESTED READING

Complete references will be found in the Bibliography at the end of the book.

Conant, Chapter 2.
Mises, pp. 205-217.
Castell, pp. 242-252.
[E], Volume 1, Book 2.

PART THREE

Problems Raised by Science

11

Determinism

> *"If any of them can explain it," said Alice, "I'll give him sixpence.* I *don't believe there is an atom of meaning in it."*
>
> *"If there is no meaning in it," said the King, "that saves a world of trouble, you know, as we needn't try to find any."*

THIS QUOTATION expresses my sentiments concerning the problem of determinism. It seems to be the most controversial pseudo-issue ever made important by dressing it up in Big Words.

METAPHYSICAL DETERMINISM

The problem is whether the future is or is not determined. I must confess that I have no idea what this means, and the more I study various papers written on this subject, the less I understand what an undetermined future would be. It is admitted, on the one hand, that the future will take place in one definite way and, on the other, that it has not yet taken place. Beyond this very little can be added.

"The Dictionary of Philosophy" gives us the following definition of the doctrine of determinism: ". . . that every fact in the universe is guided entirely by law." This is typical of the type of confusion that is necessary to deceive us into thinking that there is a problem here. We must believe that laws "guide" facts, or that laws "determine" facts. If, instead, we recognize that laws only serve as descriptions of facts, then it becomes clear that The Law of Nature describes all facts—hence that the universe is determined in this sense; it is also clear that this is a characteristic of all conceivable universes.

Another usual formulation is to ask whether every event has its cause. But we have seen that whether we describe the laws in causal form or not is entirely our free choice. Causal laws are very useful, because they lead to a division of labor. We try to formulate a law which, given the present, will determine the future for use. Then the more facts we can accumulate about the present, the more we will be able to predict concerning the future. Discussions of determinism are intermixed with discussions of our ability to find such laws, and of being able to find *all* the initial conditions.

Far be it from me to deny the importance of the question as to what degree we humans can know the future. But whether or not we humans can in principle predict the future is a question about our limitations, not about the nature of the universe. To state the fact that we can never predict the future completely in the terms "the universe is not determined" is not only bad syntax, but it leads to atrocious Metaphysics.

Let us conclude that everything relevant to this pseudo-problem has already been stated in Chapter 3, and pass on to a more interesting question.

PREDICTING THE FUTURE

Before getting lost in the problem of the predictability of the future, let us ask who is to make the predictions. Is it to be a super-human intelligence, which has unlimited knowledge of The Law of Nature, or is it to be an ideal human being, or is it to be an actual person?

On the one extreme, a super-human intelligence can "read off" the future from The Law. But this is not surprising, since The Law is a description of past, present, and future. At the other extreme, it is obvious that at the present, with our present laws, we are unable to predict the future with any accuracy. So we see that any interesting question must lie somewhere between these extremes.

We can distinguish four factors influencing the problem: (1) The laws available. (2) The facts available. (3) The reasoning powers available. (4) The time available. As to the first factor, we have already seen that we should allow ourselves less than The Law of Nature. We may suppose that the law available is a consequence of The Law—that is, describing some part of the history of the universe—or that it is an approximation to such a partial description. Instead of taking it to be a complete description of a part of the universe's history, it is more realistic to take it to be a causal law applying to certain kinds of phenomena, which, coupled with facts about the present, predicts these phenomena in the future. The problem for such laws is to know just what area can be covered by them. One school maintains that such laws will be found to cover all types of phenomena. A more conservative position is that, although this goal will never be reached, we will continually broaden the field to which such laws are applicable.

On the other side we find various views maintaining that some areas will never be describable by human laws; some of the positions will be discussed in the following chapters. Again it can be argued that, although causal laws can be found, they are always only approximately true and that, besides broadening the field of applicability, we must also try to get better and better aproximations to the truth. It is even argued frequently, especially since the successes of Quantum Mechanics, that our more precise laws cannot be causal, but must be statistical. (This last distinction is, for practical purposes, not as fundamental as it may seem, since we have already noted that due to errors in observation we can only make statistical predictions from any kind of law.) Disputes between these various positions are fruitful, and they influence significantly our beliefs as to the predictability of the future.

Let us next turn to facts. Even if two scientists agree as to what we can expect from our theories, they may disagree as to the practicability of predicting the future, on the basis of varying views as to the possibility of finding the necessary facts to apply the theory. Newton's laws will tell us all about the future of the solar system, if we know just where the planets are at a given moment in relation to the sun. Errors in prediction may be due, on the one hand, to the inaccuracy of the law (which became apparent in discrepancies in the orbit of Mercury) and, on the other hand, to inaccuracies in determining the positions of the planets, which can be reduced but never entirely eliminated. These errors "spread" in time, which is why short-range predictions are more reliable than long-range ones. It is well known to anyone who has to do a lot of calculations that the so-called rounding errors may become very significant in a long computation. Similarly a very small error in calculation may be magnified as the result is used over and over again. Introducing a small error of observation is like introducing a calculating error into the calculation of predicting. The more we use this figure, the greater the error may become. Perhaps a good analogy is provided by the taking of a drug. If the druggist makes up the prescription just slightly too strong, it may have little effect at first, but may eventually lead to death instead of a cure.

The problem of predicting is further complicated by the celebrated Uncertainty Principle. When we come to atomic and subatomic phenomena, we find that our methods of observing and measuring are so crude that we cannot help disturbing the system to be observed. When we measure the length of a table, we put a yardstick up against it, which act pushes the table. But this push is negligible. We can further reduce this push by not touching the table, but only sighting it. This, of course, involves reflecting some light from the table. This light ray also exerts a push, but so slight that no one would ever take it into account. However, when we come to measure minute particles, we are like elephants trying to measure a violet. According to the Uncertainty Principle there is an absolute limit to the accuracy we can achieve. This shows that, whether our laws are statistical or causal, there are severe limits on the accuracy of our predictions, especially long-range predictions.

The third factor, that of limitations on our powers of analysis, it not nearly as frequently mentioned. Yet there are famous examples to illustrate this difficulty. To derive predictions from a given theory often requires long and extremely difficult mathematical arguments. Given Newton's laws, it is easy to show how two bodies move relative to each other, each attracting the other one. But when we ask the same question for three bodies, the answer has still not been found in complete generality. We have all kinds of approximations for special cases, but no general solution. In the case of the solar system the motion of two planets around the sun is solved by assuming that the attraction of the planets on each other is nearly negligible compared to the pull of the sun. We first calculate the pull of the sun and the resulting planetary paths; then we introduce a correction for the mutual attraction of the planets. This method necessarily introduces new inaccuracies, which are vastly increased when we consider dozens or hundreds of billions of bodies.

The extreme example of this difficulty in calculation is given by Einstein's recent work. As we have noted, his Unified Field Theory, published in 1950, is entirely untestable for the time being, because the mathematical problems that need to be solved are beyond our ability. In this case Science must wait for major progress in Mathematics before a single prediction can be made from the new theory.

The last of the four factors, the time element, does not affect predictability in principle, but it does affect the practical problem. It is very nice to know that with our present theories we can predict tomorrow's weather. But if it takes our calculating machines a full month to carry out the necessary prediction, then it is of very little practical use. We could always argue that with the progress of machine-construction the prediction will be fast enough to be helpful. But there still remains the possibility that some predictions by their nature require more time than it takes for the event to happen. This may, for example, be the case in trying to predict human decisions.

In summary, let us recall that this problem of predictability has nothing to do with the problem of whether the universe is in some mysterious sense determined. Rather it has to do with our human abilities to comprehend the universe. At any one stage it is a scientific problem to see how far and how accurately the future is predictable by our limited means. But it is up to the philosopher to speculate which of these limitations are essential and which will be removed, or at least improved, with human progress.

TWO RELATED TOPICS

Let us first consider an ingenious argument to show that complete prediction of the future is in principle impossible. Its source is not entirely clear to me, as I have heard it attributed to various authors. Recently K. R. Popper studied this problem exhaustively.

Suppose that we have found the causal laws describing physical nature, or even that some spirit whispers them in our ears. Then we can build a machine which is specially designed to predict the future, given data concerning the present. Let us build such a machine, and put it in a laboratory which is carefully isolated from the outside. Let us wire the machine so that it will calculate the future for 100 years, or at least so that it can answer any question about the next 100 years. Suppose the machine has a certain signal indicating that the answer to our question is "Yes" and another for "No." Let us connect to the machine an electric fan, built so that it could run for hundreds of years if not shut off. Let us connect this fan so that the only way it will be shut off is by means of a "yes" signal from the machine.

Let us now suppose that we somehow supply the machine with the needed data concerning the present. It doesn't matter whether we did this by measurement, by accidentally guessing the truth, or even had a superior spirit hand it to us. Somehow the machine is given all the needed data. Then we can ask it any question concerning the next 100 years. Let us ask it: "Will the fan be running 99 years from this moment?" The machine goes into an intricate calculation, and if the future is predictable, it should sooner or later come up with an answer. If this is to be any use at all, it must arrive at an answer in less than 99 years. Suppose it arrives at the answer "Yes," telling us that the fan will be running. Then this very answer causes the fan to stop! But if the machine says "No," then the fan will keep on running for several hundred years. So in either case the machine comes up with the wrong answer.

It may occur to you that the trouble in this argument is that the machine is part of the system studied. But, alas, we too are part of the universe about which we make predictions! If we analyze the situation, we see that there is nothing wrong with the argument. So at least one of our assumptions must be wrong, and we see that these correspond to the four factors affecting predictions. First, we assumed that we somehow have the laws governing such physical problems. Perhaps these laws cannot be made comprehensible to us, and hence to our machines. Secondly, we assumed that somehow we had all the initial data necessary available to us. Perhaps this is beyond us. Thirdly, we assumed that the machine can carry out the necessary calculation, while it may be the case that such problems prove unsolvable to it. Finally, and perhaps most simply, it may be the case that it always takes the machine more than 99 years to finish the calculation, in which case the contradiction is removed. The machine sends its "yes" answer in the hundredth year. This turns the fan off, but it does not alter the fact that the fan was still running after 99 years. Any one of these factors or all of them could make the required prediction impossible, and it would be most interesting to determine just which is the actual cause.

The second topic relates specifically to the difficulty in solving the mathematical problems involved in predictions. K. Gödel proved a most interesting theorem to the effect that no matter what mathematical methods we have at our disposal, some of the questions we can ask ourselves cannot be answered by these methods. This does not preclude that with the progress of Mathematics a given unsolvable problem will be solved; however, then there will be new unanswerable questions. In other words, the progress of Mathematics is an infinite progress, where at each stage there remain unanswerable questions.

We know that each body of theories in Science is an interpreted mathematical system of a fairly advanced type. Hence it has some unanswerable questions in it. What we don't know is whether any of these unanswerable questions concern predictions. But we have no reason to suppose that none of them do. It is entirely possible that even if we knew exact causal laws applying to all types of phenomena, and if we had an unlimited supply of data, we would still have to wait for mathematical progress for certain predictions. For some reason this possibility seems to have been generally overlooked. When we find some Physics problem as yet unanswered, we assume that it is because no one has yet found the right "trick" to solve it. Actually it is possible that this problem needs an entirely new branch of Mathematics.

This suggests the interesting possibility that, not only must we expect an endless advance on the part of the theoretical scientists, but that this progress must go hand in hand with the endless progress of the mathematicians, if it is to be of use to mankind.

SUGGESTED READING

Complete references will be found in the Bibliography at the end of the book.

Castell, pp. 112-115.
*Hume, Section VIII.
Frank [2], Chapter 9.
*Eddington, Chapter XIV.
Planck, Section 7.

12

Life

"If you think we are wax-works," Tweedledum said, "you ought to pay, you know. Wax-works weren't made to be looked at for nothing. Nohow!"
"Contrariwise," added the one marked 'DEE,' "if you think we're alive, you ought to speak."

SCIENCE'S GREATEST SUCCESSES come mostly from inanimate nature. When I had occasion to give examples of scientific theories, I almost always turned to Physics, the parent of the other sciences. I must now ask what Science can say about Life, and I will immediately find myself in the midst of several heated disputes.

THE LIMITATIONS OF SCIENCE

We are accustomed to associate the word "Science" with telephones, atoms, strange drugs, rockets, or even interplanetary television. We are rarely surprised at any prediction of what future Science will do—as long as it deals with inanimate matter. Most of us have strong prejudices as to the limitations of Science when it treats living organisms.

Of course we realize that Science has made some progress in the study of Life. But we tend to think of this as dealing mainly with low forms and as becoming less significant as we climb the evolutionary scale. Under no circumstances is this progress comparable to the triumphs of Physics. We are happy if scientists can predict eclipses of the moon millennia ahead, and we are not too disturbed if a scientist is fairly successful in predicting the colors of the eyes of as yet unborn fruit flies. But we revolt at the idea of having a scientist predict our future actions.

No doubt it is an accurate evaluation of the present status of Science to say that Physics and Chemistry dwarf Biology and the Social Sciences. The issue, however, is the future potential of these underdeveloped fields. Here we run into violent arguments, strong prejudices, and very little clear thinking. It is a topic on which a debate appears to be impossible: A debate requires two debaters on each side, and on this issue no two people seem to have the same position.

But as a rough first step we may classify our partisans into two camps, the vitalists and the mechanists. For a definition of these terms we will turn to *The Dictionary of Philosophy,* where we find "vitalism" defined: "The doctrine that phenomena of life possess a character sui generis by virtue of which they differ radically from physico-chemical phenomena. The vitalist ascribes the activities of living organisms to the operation of a 'vital force' such as Driesch's 'entelechy' or Bergson's 'élan vital.' Opposed to vitalism is biological mechanism which asserts that living phenomena can be explained exclusively in physico-chemical terms."

This definition is a good start for our considerations. But it contains many Big Words, badly in need of explication. It will be good to take up these concepts in terms of a concrete example. For this we turn to the Theory of Evolution.

EVOLUTION

For the sake of illustration we will take up several conflicting theories of how living organisms evolved.

Perhaps the earliest theory in western culture is furnished by the Bible. It is one of the only too recent successes of scientific development to convince the man on the street that there is no religious danger in rejecting the scientific content of the Bible. It took two millennia for mankind to learn that the moral teachings of the Bible are independent of its pseudo-scientific content, just as the logical content of *Principia Mathematica* does not depend for its validity on Mr. Bertrand Russell's personal moral convictions.

The essential point in the Bible's account that was disproved by the weight of evidence is that various species were created independently of each other. In the face of overwhelming evidence, we are forced to accept the theory that species evolved from simple to complex forms, through—perhaps—a billion years. This one doctrine is common to all current evolutionary theories. I cannot help remark that this account of the development of Life appears to me much more awe-inspiring than the biblical version.

Let me first give a brief account of that part of the various theories which is quite generally accepted. Most of what I have to say was known to Darwin. I merely add the mechanism of Mendel's laws and of the theory of mutations developed by de Vries.

Perhaps the most fundamental fact of living beings is their ability to reproduce, to create other living beings in more or less their own image. And they can produce a fabulous number of these. The human race is one of the slowest to reproduce; yet given a man and a woman, they could certainly have ten children by the age of forty, which in turn could reproduce at the same rate without too much difficulty. If Adam and Eve reproduced at this rate, and if their children all lived to do the same, and their grandchildren as well, then in less than a thousand years there would be so many human beings that even if they stood elbow to elbow they would not have room to stand on the surface of the earth. As a matter of fact, in a thousand years they would have to have over a hundred people standing on top of each other to have room on the earth. These simple figures prove conclusively that there is indeed a struggle for existence. Add to this all the other species, some of which reproduce much more rapidly, and we see that every generation of every species must fight to find room to live in and for the means of existence.

This struggle is not caused by any malicious desire on the part of some individuals to exterminate others, but by the physical necessity of killing other life in order to have room to live in and food to subsist on. In this struggle those individuals having some special qualification will be more likely to survive, and soon their descendants will populate the earth. But how do individuals happen to have special qualifications?

According to Mendel's laws, the new generation has the traits of its parents, only reshuffled. The genes which carry the pattern of life seem to be immortal, and they are merely combined in new ways in succeeding generations. If this were all Nature provided for, we might see the extinction of some known traits, but new species would be possible only by the recombination of traits. Fortunately, once in a long while a gene goes out of order, it changes into something entirely different. Generally, perhaps in 99 per cent of all cases, this means a harmful change leading to the death of the infant carrying this *mutation*. But the hundredth case will endow its possessor with an advantage over his competitors, which will slowly but certainly allow him to become master.

Suppose that on an island there are 1000 horses, more or less alike. The island has but a limited supply of grass, and hence the number of horses stays about the same. There are also some horse-eating tigers on the island, which constantly threaten the existence of the race. One day a young colt is born with a mutation of genes, enabling him to run faster. He is lucky enough to survive, and in the next few generations we find that we now have ten horses of the fleet variety and 990 "normals." Say they each have about ten colts, nine of which die in the struggle for food, keeping the total number constant. (I am ignoring the sex of the horses for simplicity.) But the fleet of foot colts have a slightly better chance of escaping being eaten by the tigers, and hence in the next generation we find 12 of these, and perhaps 13 in the one after that, and so on—slight increases varying with setbacks. The laws of probability tell us that even though they will remain exceptional for several generations, we can expect that in a few thousand generations the fleet of foot horses will entirely replace the "normal" variety.

An ingenious mathematical theory describing such a situation can be based on a gambling model. We may think of the two types of horses as gambling against each other. The stakes are room to live in, where in each generation there is a small change—room for one horse is lost by one group to the other. The fleet of foot subspecies has a slight advantage, say they have a 55 per cent chance of making a gain, as opposed to a 45 per cent chance for the opponents. The theory then predicts that eventually one "gambler" will wipe out his opponent. The prediction favors the fleet of foot horses by a nearly 7:1 margin, in spite of their initially unfavorable position. And the prediction is that it will take between 8000 and 9000 generations for one of the subspecies to disappear. Naturally, this is a very crude model, but much more realistic models can also be stated in this very convenient form.

If we add up the effects of all mutations and of selection acting on a scale of a billion years, we are supposed to have an explanation of the history of Evolution. This is the theory of neo-Darwinism, or Darwin's theory up to date. We are told that these mutations, occurring quite at random, combined with the machinery of natural selection acting over so long a time, can explain all the fascinating variety of living forms.

The great difficulty in evaluating this theory lies in its incompleteness. It is more of a qualitative description than a precise scientific theory. The proponents of neo-Darwinism claim that there is no known instance of evolution which they cannot explain. This is actually untrue. What is true is that no such instance clearly contradicts their theory, but this is not surprising when we realize how little the theory actually states. To say that the known changes *could* have been brought about by the described machinery does not explain these changes. We have seen that an adequate explanation is one which would have enabled us to predict the outcome, before it took place. But none of the present evolutionary theories enable us to make such predictions.

The most famous example of a series of evolutionary changes is the development of the modern horse. Fifty million years ago there was a creature roaming on the earth which we now believe to be the earliest recognizable ancestor of our familiar horses. This remote ancestor had four toes, and fed on the juicy leaves of low trees. Through the ages its size increased fairly steadily, its toes changed first to three and finally to one, and its teeth gradually evolved until now they are perfectly suited for grazing instead of browsing. On the basis of natural selection we must look for an explanation somewhat like this: It is to the advantage of any animal (large enough not to be able to hide) to increase in size, to give it a better "fighting chance" for survival. For example, as its legs grow, its chances of outrunning its most dangerous foes increase. The change in number of toes also helps it to run better The change in teeth is brought about by a change in environment. As it becomes more difficult to acquire a sufficient supply of leaves from trees, the horses (or rather their ancestors) had to turn to a new supply of food, grass. Their present teeth are like miniature millstones, excellently suited for grinding up grass. As all these changes make it easier for the individuals to survive, in the long struggle for existence the individuals with mutations tending in the advantageous directions win out over the normal members of the species, and slowly but steadily the species adapts itself better and better to the environment, until the present horse evolves. (For some reason the accounts usually give us the impression that we have reached the end of the horse's development.)

There is no doubt that the horse *could* have evolved in the manner described. But had Mr. Darwin lived fifty million years ago, he would certainly not have been able to predict that these changes would occur, even if he had known how the environment was going to change. Since his theory would not have served for predictions then, it is not adequate for an explanation now. It is a scheme of explanation, but the details are yet to be filled in.

THE OPPOSITION

One of the most vocal opposition groups is the school of Lamarckians. Lamarck's theory of evolution precedes that of Darwin; as a matter of fact, his most significant work appeared the year of Darwin's birth, 1809. The part of his theory that must concern us is his belief that characteristics acquired during one's life can be inherited. This is the way in which he tried to account for the way species adapt themselves to their environment: The parent generation feels the need for a certain change—for example, stronger muscles—and brings this about by continual use of these muscles, as a result of which the offspring are born with somewhat better muscles, and slowly a great change comes about.

Darwin himself subscribed to this theory, but the neo-Darwinists abandoned it in favor of random mutations. The great difficulty in believing this theory is that we see no way in which acquired changes could be passed on to the next generation. We would have to believe that by using our arms we not only strengthen them, but change the gene structure of our reproductive cells. Beside this theoretical difficulty, there have also been experiments designed to test Lamarck's theory, and although I do not consider them conclusive, they do make the theory much less credible than an acceptable theory must be.

Why is it then that Lamarck's theory has not been entirely abandoned? (For example, it happens to be the official theory of the Soviet Union.) Because many biologists question whether neo-Darwinism can account for the rapidity of evolution. In an incomplete qualitative theory it is very difficult even to estimate how rapidly the processes should progress, but some estimates have been made, and they are disturbing. According to the theory of probability it is to be expected (granted neo-Darwinism) that Evolution should have taken place, but some estimates seem to show that it should have taken longer for nature to get as far as it did.

As an example let us consider the development of the species *Homo sapiens*. We believe that his ancestors of a million years ago were hardly recognizable as potentially human. How many mutations took place during that relatively short period, and how many of the rare favorable mutations were needed to bring about Modern Man? There were about 40,000 generations in that period. We have historic evidence of the fact that the abundance of the human race has taken place in historic time, and that at the dawn of history Man numbered in the tens of thousands. Thus we may be safe in assuming that the total number of human-like beings since the days of the "missing link" was only about the same as the number alive at the present moment! We should, in accordance with neo-Darwinism, assume that there are enough mutations present on the surface of the earth right now for a machinery as simple as natural selection to combine the existing traits into a superhuman species. Although this is not beyond the realm of imagination, it does sound somewhat unlikely. This is the kind of argument that sheds doubt on whether neo-Darwinism suffices to account for the rapidity of the history of Evolution.

If this sort of argument is reasonable, and we will know this only when neo-Darwinism is made more precise, then there must have been factors which brought about mutations (or other changes in the chromosome-structure) in a nonrandom way. If the environment could directly, or through the activities of the individual indirectly, bring about inheritable changes which adapt the animal better to its environment, then the time-scale appears less surprising. In this case we find that there will be a surplus of favorable mutations, which would speed up the entire process.

Let us say, cautiously, that although Lamarckianism has been discredited, we have not heard the last word in the dispute. Even if their positive statements are wrong, the Lamarckians may be quite right in their criticism of neo-Darwinism. We will not be able to check this latter theory conclusively until it is completed to the point where it can really explain, at which time we will see whether it accounts for the speed of Evolution.

So far we have discussed the two best-known mechanistic explanations. We must now turn to their two vitalistic rivals. The claim will be that there are certain specifically living forces, which are not to be found in inanimate nature; only by taking these properly into account can we account for all the facts of Evolution.

First of all we must note that there seems to be a strange tendency in evolution for "straight-line" development. The size of horses (and of many other species) keeps increasing through the ages. The teeth of the saber-toothed tigers increase gradually, until they are no longer teeth, but knives. The various lines in the family of Titanotheriidae all develop V-shaped horns, and grow to enormous sizes. Many species become more and more "fancy" until they are finally extinguished. And so on.

We must take care not to get involved in useless arguments as to whether such lines of development are really straight. I would like to illustrate this by showing that even very good men can go astray. Simpson, in his very interesting book, *Meaning of Evolution,* makes up a hypothetical example of supposed straight-line evolution and discusses it at length. I will draw a figure somewhat smiliar to his, and draw in the curve that he considers most reasonable. It is certainly true that this is the most credible curve, but is this really proof that the evolution did not proceed in a straight line? Let us study both axes carefully.

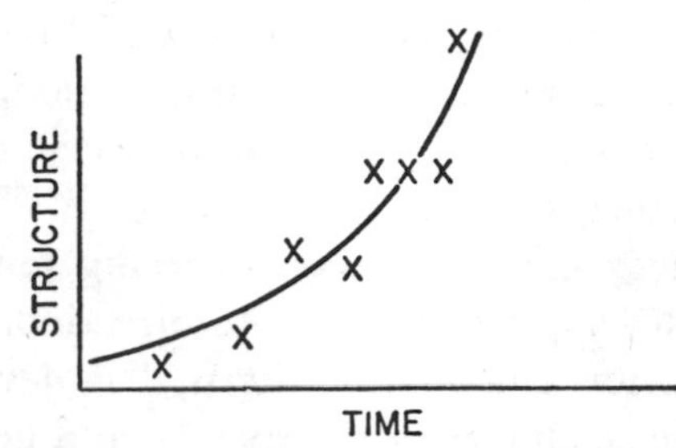

How does one measure "structure"? Do we really have a good measure of it in terms of numbers? If it is only qualitatively measurable, then how can we plot it? And even if we have a numerical measure, such as the size of the animal, there is a great deal of arbitrariness left. We could use its length from head to tail, or an average dimension, or its volume, or its weight. The curve may be straight according to one measure, and not according to the others. And who is to say that the theoretically significant measure of structure is not something more complex, like the logarithm of the volume? There are many precedents for this in Physics. But even if all these questions are successfully answered, there are unanswerable questions concerning the time-axis. To show this, I will redraw the same data, assuming that the same structures occurred, in the same order, but at somewhat different times.

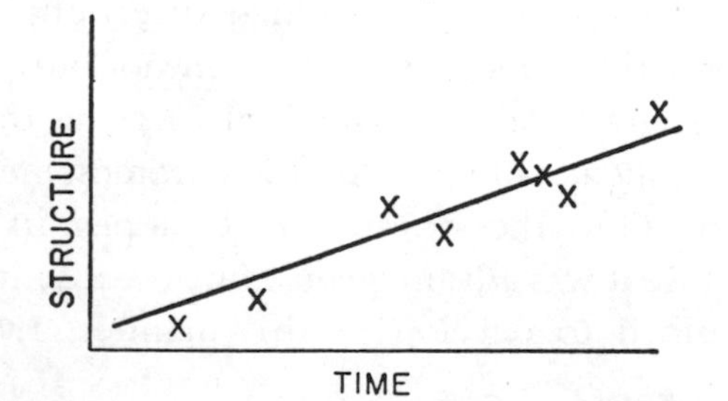

Now the points definitely lie on a straight line. But this does not make sense! How is it possible that the straightness of the succession of structures depends on the frequency of changes? The same changes occurring at one rate are straight, at another are not. Clearly, this indicates that the concept of straightness is not reasonable. What one means by straight lines in evolution is not something that can be plotted as a straight line on a graph paper, but a steady tendency in a given direction. This is a good example of the danger of confusion between qualitative and quantitative arguments.

How can Biology explain this unwavering tendency in some species to keep developing in the same direction? The only possible explanation for a neo-Darwinist is that developing in this direction is to the advantage of the species, and hence the mechanism of selection brings about changes in this direction. But in many cases we must strain our imagination to see any possible way in which these continued changes helped the species. In many instances the early steps in this direction were a definite advantage, but the later ones seem detrimental and even seem to be the main cause of the extinction of the species. Growing in size is an advantage in combat, but too great a bulk makes it impossible for the individual to find enough food for survival. Similarly it has been argued that the too long teeth of the sabertooth and the enormous horns of the titanotheres were more of a handicap than an aid. Also, it is very difficult to see how some of the fancy growths of species (generally found shortly before extinction) could have been of any imaginable use.

The school making most use of these arguments is the one believing in orthogenesis, that is, in the inherent tendency of living organisms to develop in predetermined directions. We are told that the reproductive cells contain in them not only the means of determining the traits of the individual carrying them, but also certain innate tendencies to develop the chromosomes in a certain fixed direction. Thus the sabertooth developed its name-giving organ not because it was advantageous, but because its protoplasm was predetermined to develop in this manner. Evolution as a whole is the sum total of such straight-line developments, according to this vitalistic school.

It is questionable whether orthogenesis is a theory at all. It is much too vaguely stated to have any explanatory value. What prediction could we make from such straight-line tendencies, unless we had a clear-cut definition of what such a tendency is? I take orthogenesis rather to be a criticism of the currently accepted theory, pointing out that the present theory (neo-Darwinism) cannot account adequately for these developments. Orthogenesis is hardly a general theory, since any unprejudiced observer must recognize the vast number of cases in which no steady tendency is shown by a species. But it does point out certain, so far unexplained trends, and it states the hypothesis that these trends cannot be explained by a purely mechanistic theory. It maintains that there is some inherent trait of living organisms (not present in inanimate nature) which brings these developments about, though they do not state (except lyrically) what it is.

The last of the four theories to be discussed here is distinguished by having had two of our greatest literary geniuses advocate it: Bergson and Shaw. Shaw had his "life force" and Bergson his "élan vital," inherent forces in life, unconscious striving on the part of living beings to perfect themselves. These forces are the causes of Evolution. The remarks that I made about orthogenesis apply here just as well, with the notable difference that Mr. Shaw's "Back to Methuselah" is by far the most delightful reading matter in the evolutionary literature.

In conclusion, we find a great deal of agreement, and one fundamental disagreement. It is agreed that there is a natural surplus of life, creating the celebrated struggle, and hence a natural selective process preserves the individuals showing minute advantages over their brothers. These differences are brought about by changes in the chromosomes, or mutations. But just what causes these changes is open for dispute. We are told that they are random changes, that use and disuse affect the genes, or that they manifest an inherent trend in a given direction or a life-urge, both peculiar to living beings. The hypothesis of random changes is simplest and is the accepted theory of the day. But it is far from being a complete quantitative theory, and its critics feel that, if it is made more precise, it will prove to be inadequate. In this case it will have to be supplemented by auxiliary hypotheses. Some such hypotheses have already appeared—for example, that certain organs, like horns, grow in proportion to size; hence the explanation of the enormous horns of certain species is that growth of the animal was biologically useful, and growth of the horns was an accidental by-product. The Lamarckian hypothesis could be another such mechanistic extension of neo-Darwinism. But two schools maintain that the theory can be made adequate only by the introduction of certain typically life-features, whatever these may turn out to be. This contention is at the root of the mechanist vs. vitalist dispute.

CRITERIA FOR AN ADEQUATE THEORY OF EVOLUTION

Before we take up the basic dispute, consider what changes must take place to make the Theory of Evolution an adequate scientific theory.

Let us return to the very simple gambling model of evolution and ask whether this describes neo-Darwinism correctly. This is a question unanswerable at the present state of the theory. For example, the model requires us to state how much of an advantage a favorable mutation provides for an animal. We picked the arbitrary number of 55 per cent chance of winning a place from the nonmutants. But there is nothing in the theory that would provide us with such a number. It is even more questionable as to whether in a fixed population situation it is reasonable to assume that in each generation one or any other number of animals will change from one type to the other. There is nothing in the theory that would describe specifically the way in which one type competes with another type.

We would need details on the points just discussed. We also need data on the percentage of favorable mutations, and on the percentage of these which in themselves make an outwardly significant change in the individual. Naturally, we need data on the frequency of any mutation taking place.

To go even a step further down, we know relatively little of the way genes control traits. We know that for a given genetic structure it is possible to alter the outward appearance of the individual by various biochemical means. To what extent does the environment interact with the genes in producing physical characteristics?

All these and many more questions will have to be answered before an evolutionary theory emerges that can make even simple predictions. But there is hope that recent experimental advances, especially the ability to produce mutations in large numbers by means of radiation, will hasten the development of this field.

REDUCTION OF A BRANCH OF SCIENCE

Roughly speaking, a mechanist maintains that Biology can be reduced to a branch of Physics (and Chemistry). Just what this could mean is our problem in the present section.

It will be convenient here to speak of Physics as if it included Chemistry, which, after all, has almost become a branch of Physics. And this is one of the famous examples of reduction. We have seen (in Chapter 10) that Science as a whole is defined by a common method, and that it is divided into branches by the types of phenomena to which this method is applied. For example, one branch of Science deals with heat phenomena, another (Mechanics) with the motion of particles. (The historical accident that these are both classified under the heading "Physics" is of no interest to us at the moment.) When we play pool, the motion of the various balls is of interest in Mechanics, while the heating of the green cover by continued friction is a heat phenomenon. Slowly theories build up in both branches, introducing whatever theoretical terms seem most suitable for their particular field of interest. The two branches are distinct if the terms in one are not exactly the same as the terms of the other. For example, "kinetic energy" is peculiar to Mechanics, while "temperature" is used in the study of heat. It may turn out, however, that we can find a theory in purely mechanical terms which accounts for all the heat phenomena. In that case we say that the latter branch is *reduced* to Mechanics.

This did actually occur, at least for gases. (For solids we must take Quantum Mechanics into account, which complicates the picture.) It was found that the temperature of a gas is proportional to the average kinetic energy of the molecules of the gas, which means that all theories about temperatures can be expressed in terms of the speeds of the molecules. For example, the statement "the gas becomes hotter" expresses the same fact as "the velocities of the molecules of the gas have, on the average, increased."

This way Science can reduce the number of different theoretical terms it needs, and hence simplify the body of theories. This is always the sign of genuine scientific progress. There are many such examples in Science. The various laws of Kepler, Galileo, and of the tides can all be reduced to the branch studying gravitation by means of Newton's laws. Most of Chemistry was reduced to Physics when we found that the valence of an element bore a simple relation to the number of electrons in the outer orbit of the atom. These reductions simplify and unify Science, and are therefore very important in the history of Science (see Chapter 10).

We must now try to find just what a reduction accomplishes. Let us talk of a first branch and a second one, with the second to be reduced to the first one. Each will have its own theoretical terms and its characteristic phenomena. We will say that *the second branch is reducible to the first one if there is an acceptable theory in terms of the theoretical concepts of the first branch which explains the phenomena of the second branch.*

This definition is very important, and I must clarify some points. Whether a branch is reducible may depend on our progress of knowledge, because the reducing theory may not become credible until further facts are discovered. For example, the theory concerning the number of electrons, and the remaining atomic theories needed for the reduction of Chemistry, would not have been credible without the vast new data supplied by our twentieth-century instruments. When this theory is finally found and accepted, we can say that the second branch has been reduced. It is understood, of course, that acceptability is defined in terms of the credibility of the theory.

This leaves the troublesome point of having to say that the theory "explains the phenomena of the second branch." I do not mean that all phenomena are explained. As a matter of fact this is not possible. From the scheme of explanations we know that some phenomena must be used (in addition to the theories) to explain the remaining phenomena. Even then, it is only in a Utopian Science that we can hope to explain all the remaining phenomena. We should rather interpret the phrase as requiring that the theory is as good in explaining phenomena of the second type as can be expected; for example, at least as good as the theories of the second branch were. In other words, we require that this new theory represent progress, even if we then drop all the theories of the second branch. Only then can we feel that this branch has been reduced, or eliminated as a separate branch of Science.

There is a way of strengthening the requirements for reduction and a way of weakening them. We could require not only that there be a reducing theory, but that this theory be already known at the present time; this would be the sense of reducing the second branch to the first branch *in its present form*. It differs from the definition above in that it does not allow improvements in the reducing theories. We could go to the other extreme and allow not only improvement in theories, but also in the theoretical vocabulary of the first branch, in which case we would have reduction to the first branch *in the extended form*.

We will now apply these considerations to the dispute between mechanists and vitalists. In this case, the first branch of Science is Physics, the second Biology. We can ask in any of the three senses whether Biology is reducible to Physics. If we require reduction to the present form of Physics, we have the extreme mechanist position which I shall denote by M_1. If we require ordinary reduction, we have the more tolerant position M_2, and if we allow improvements in the theoretical concepts of Physics, we have the mechanist position M_3. The denial of these gives three possible vitalist positions, increasing in strength.

But there are more extreme vitalist positions. A vitalist could assert not only that reduction in any of these senses is impossible, but that we *must* introduce a "life force" or a "life substance," like Driesch's very controversial "entelechy." A man might call himself a mechanist just because he denies the existence of such a new force or substance, a position denoted by M_4. Finally, there is the very extreme vitalist, discussed before, who denies that Biology will ever become a real branch of Science, who holds that the laws governing life are beyond human comprehension. Hence a scientist might classify himself as a mechanist just because he denies this extreme case; this weakest possible mechanist is M_5. The denial of these five positions give us five possible positions for a vitalist, which I shall call V_1 to V_5, respectively.

By consulting the chart of possible positions we see that it is entirely possible for someone to be a mechanist in one sense and to be a vitalist in another. Someone could hold that Biology is reducible to an improved Physics (and hence hold M_3), but deny that the present terminology is adequate for reduction (that is, deny M_2, which is the position V_2). In general, each new mechanist position is compatible with the denial of all the previous ones; and, if someone tells us that he adheres to M_3, he must still tell us how he feels about M_1 and M_2. Hence, any man who has completely made up his mind about this controversy must hold one of six positions, which are also shown in the chart below.

POSSIBLE MECHANIST AND VITALIST POSITIONS

STRONGER MECHANISTIC POSITIONS ↑				STRONGER VITALISTIC POSITIONS ↓
M_1	Biology is reducible to present-day Physics.	Present-day Physics is not adequate for reduction.	V_1	
M_2	The theories of Physics can be strengthened to give a reduction.	A strengthening of theories is not enough to allow reduction to Physics.	V_2	
M_3	An improvement of the vocabulary and theories of Physics will result in reduction.	Even a change in vocabulary cannot enable Physics to reduce.	V_3	
M_4	There is no need for introducing a "life force" or any new substance.	Any adequate body of Biological theories must refer to a new force or substance.	V_4	
M_5	The laws of Biology are within human reach.	Biology will never become a branch of Science.	V_5	

MORE MECHANISTIC ↑

P_1 (M_1)	:	Biology is reducible to present-day Physics.
P_2 (M_2 and V_1):		Present-day Physics will not serve for reduction, but all we need is progress in the theories.
P_3 (M_3 and V_2):		There is no reducing theory in terms of the present day physical vocabulary, but with progress of Physics such a theory (in new terms) will be found.
P_4 (M_4 and V_3):		Biology is not reducible to Physics, but there is no need for a "life force" or "life substance."
P_5 (M_5 and V_4):		A science of Biology can be developed, if one allows a "life force" or "life substance."
P_6 (V_5)	:	No science of Biology is humanly possible.

MORE VITALISTIC ↓

SIX COMPLETELY SPECIFIED POSITIONS

We have carefully considered the relative claims of the two schools and have arrived at a clarification of the various possible positions people can hold. Some of these are more frequent than others, but all six have at one time been seriously defended. This sort of clarification is called logical analysis, and I believe that the most important progress in Philosophy in the near future will come from such analysis. It requires careful acquaintance with modern tools of Logic and a great deal of patience; but the reward is great: We can finally sink our teeth into a clear problem and can escape all the frustration of not really knowing what we are arguing about.

Now that the problem is clear, we must try to see what we can say about the solution.

MECHANISM VS. VITALISM

Of course, any of the six positions may turn out to be correct, but they are not equally likely. It will be best to consider them in pairs.

Let us first take the extreme positions P_1 and P_6. I believe that at least these are sure to be wrong. The most extreme mechanistic faction P_1 holds that all phenomena of life are explainable by means of our present body of physical and chemical theories. The reason I feel sure that this is not true is that these theories do not seem adequate even for inanimate phenomena. Most physicists will agree that our present theories do not suffice to understand the nucleus of the atom, for example, and that the theories will have to develop considerably before the fundamental atomic processes are explained. There is also the fact that so far Physics and Chemistry have been able to do very little for Biology. This last fact is constantly cited by vitalists as evidence for their views, but it may only indicate that these theories must further develop before they make their contribution to the study of life. So we see that the most extreme mechanist position, that our present Physics and Chemistry will prove to suffice for the study of life, is most likely to be false.

The extreme vitalist position, P_6, is that Science will never be able to understand living processes. Sometimes this is stated as asserting that living creatures do not obey exact laws. This we know is senseless, since there are laws always for any process. But it may be reformulated as the belief that the laws describing the behavior of living beings are too complex for human comprehension. This may be true, but I for one refuse to believe it. Science has already learned to explain simple living processes, and I see no reason why we should not progress indefinitely.

Both of these extreme positions are dangerous. Oddly enough, they have a very similar stagnating effect: They stop us from looking for biological laws. The radical mechanist is satisfied with the existing theories, while the radical vitalist denies that there is anything we could possibly find. But in my opinion the radical vitalist is by far the more dangerous of the two; if the mechanist fails for several centuries, he will eventually have to admit that there is need for improvement, but the vitalist urges us to give up all hope for explaining life scientifically once and for all. There is no scientific way of justifying this skepticism. It is a philosophical prejudice, based on the mistaken belief that if Science can understand Man, then this reduces his dignity by putting him on par with a stone. I strongly believe that this is the kind of Philosophy which, if successful, has the worst possible effect on Science.

Fortunately, these radical vitalists are not likely to have much effect outside their own classrooms.

All the other theories have an intermediate status. P_2 and P_5 are fairly clearly mechanistic and vitalistic, respectively, but extremists may not agree to this classification. These positions, though not as unlikely as P_1 and P_6, still have relatively few advocates. P_2 commits us to the view that the present conceptual structure of Physics will suffice for Biology. Yet many scientists would hold that this structure will not even suffice for Physics. It would be very strange if, after a few centuries of rapid change in Physics, we would suddenly have reached the end of fundamentally new ideas in the field.

It is position P_5 that has given Vitalism a very bad name. The advocates, like Driesch (the inventor of "entelechy"), are accused of introducing a mysterious new force or substance, just to hide their ignorance. It is true that when they are asked for a distinguishing characteristic of entelechy, all we are told is that entelechy is what differentiates life from nonlife. This sounds very much like the explanation of why morphine puts us to sleep (see Chapter 9), which in the last analysis says that it puts us to sleep because it has the power of putting us to sleep. Ernest Nagel, whose interesting article on this topic I have made use of in my analysis, states that the main reason this type of Vitalism is almost dead is that it has proved completely infertile for Science. This is probably the main psychological factor, but there are sound methodological reasons for skepticism too: Introducing a new term where the need for new terms has not been demonstrated is indefensible, and whatever value it could have had is lost by its absolutely vague status.

We are left with P_3 and P_4, the "middle" positions. Both agree that Biology will become a science, and that this will require the addition of new terms to those of the Physics of today. But can we achieve this with new physcial terms, or must we use specifically biological ones? This, to me, is the most reasonable version of the dispute. Nagel states the problem as one of trying to explain the behavior of organisms in terms of the behavior of its parts, knowing how the organism is built up of these parts. A human being is built up of various molecules in a certain definite manner. The behavior of molecules is governed by physical (or chemical) laws. Can we from these deduce the laws of the human being? I have already allowed improvements of the physical laws, even more fruitful basic terms. But there is a limit to how far one can allow this improvement, if the problem is not to lose meaning.

Suppose I am trying to explain why such an organism can reproduce. Am I to allow the law "molecules combined in just such a manner can reproduce" as a law of Physics? Certainly not! The phrase "combined in just such a manner" surely describes a biological property, and hence we achieve reduction to Physics only at the cost of making Biology a part of Physics by a trick. The analogy to this trick would be the following: Suppose someone argues that there is no way of translating the number system into the language of a certain primitive tribe (some of them have no way of naming numbers greater than 3). I could "prove" this wrong by saying "three-and-one," "three-and-one-and-one," etc. But then I have really changed the primitive language, precisely by adding the number system, even if in a very crude form.

Nagel is no doubt correct in saying that it is a belief that wholes have properties not explainable in terms of their parts (whatever this may mean precisely) which motivates many vitalists. But I believe that the final decision will have to be made as to whether biological theories will always require terms which we cannot call physical, or in other words, whether P_3 or P_4 is correct.

In many cases this is extremely difficult to decide. The term "mechanism" has a history that is most instructive. René Descartes is probably the most celebrated advocate of this theory, and it was his belief that all of Science can be reduced to the terms of Physics. It so happens that in his day Mechanics was the only branch of Physics with a respectable status, and hence he stated the thesis as reducibility to Mechanics. Since then Physics has vastly expanded, but we are told that the *same* problem of reducibility of all of Science to Physics still holds. But this is hardly the same problem, because Physics is no longer identified with Mechanics. For example, electromagnetic theory, which scientists first tried to reduce to Mechanics, is now recognized as an independent branch, and it too is available for reduction to Physics. If the trend continues, and every really well-developed body of theories is admitted into Physics, then by definition P_3 will follow rather than P_4—because, no matter what new concepts we introduce, we will call it Physics if it succeeds.

We might try to say that a small change is allowed, but no radical change in terminology should be admitted into the theories. I will cite two famous cases to show that this is not acceptable. General Relativity (specifically, its gravitational theory) took the place of Newton's theories. It covers roughly the same field, Mechanics and the law of gravitation, but the concepts are radically new. There is no use made of forces any more; instead the theory is a law specifying the geometry of the universe. On the other hand, Einstein's last theory, Unified Field Theory, unites these branches with electromagnetic theory (we do not know yet whether successfully, but that does not matter for the example), and yet its language differs from the older theory only in a slight mathematical modification: instead of symmetric tensors it allows nonsymmetric tensors. This extension is too mathematical to be called "introducing new terms," and yet it allows the adding of a new branch of Science. How are we to draw the border line?

Perhaps the difference is a pragmatic one—not one of content but of the best way of finding the theory. According to the mechanists, if we study inanimate matter (which is simpler to work with) long enough and with sufficient care, we will find a theory adequate for all phenomena. The vitalist, on the other hand, would maintain that as long as we think of living beings as if they were machines, the true laws will evade us. In this sense, and only in this one, I would side with the vitalists. It is better to follow both approaches, not only that of the physicist. As a matter of fact it must be noted that the few theories we have in Biology today, which includes all four evolutionary theories, all contain terms not taken from the theories of Physics.

Perhaps the past points to the following solution as most likely: We will find a new theory, covering both fields, in new terms; and inanimate nature will appear as the simplest extreme case of this law. The present laws of Physics will then be deducible, perhaps in improved form, as applying to such simple cases. This is the relation that existed, for example, between Newton's theory and Galileo's theory which applied only to free fall. Should this come about, I would be tempted to say that Physics was reduced to Biology and not Biology to Physics. I am not at all certain that this will be the case, but I feel quite certain that, whatever the reducing theory may turn out to be, both Physics and Biology will claim it as its own, and the great dispute will end happily with both sides convinced that they were right all the time.

SUGGESTED READING

Complete references will be found in the Bibliography at the end of the book.

Evolution.
- Darwin.
- Simpson, Chapters X, XI.
- Sullivan, pp. 78-96.
- Shaw, Preface.

Vitalism.
- Cassirer, Chapter XI.
- Nagel [1].
- Werkmeister, pp. 317-365.

Reduction.
- Kemeny and Oppenheim.

13

The Mind

"Why you're only a sort of thing in his dream!" (said Tweedledee.)

"If that there King was to wake," added Tweedledum, "you'd go out—bang!—just like a candle!"

WE HAVE FINISHED the difficult question of what—if anything—distinguishes the living from the inanimate. Now we find that our troubles have just begun: we have to ask what distinguishes "mere" living things from those higher creatures that think.

THE TRADITIONAL DIVISIONS

We may divide the usual philosophical answers into two main categories: (1) Those which hold that there is no fundamental difference between mind and matter, the monistic answers. (2) Those that assert a basic distinction, the dualistic answers.

Let us first consider monistic solutions. If mind and matter are not *really* different, then either mind is some special form of matter (for example, highly organized matter acting in a certain way), or matter is a by-product of mind (as Alice is part of the King's dream), or there is a fundamental substance of which they are two different aspects. Bertrand Russell holds the latter view, maintaining that if you look at events in one way they appear to be material, from another point of view they are mental.

On the other hand, if you are a dualist, you hold that the two are entirely different sorts of substances. You may hold that they are connected by causal relations, or even that they are entirely distinct, but synchronized like two clocks.

This gives us five possible points of view, the two dualist positions, and the three monistic ones, materialism, idealism, and neutral monism. Needless to say, all five of these positions have been held at one time or another, and I believe that with some care we could classify all philosophers into one of these five schools, though most of them would object to this, pointing out subtle differences which distinguish them from all other thinkers.

There are difficulties peculiar to each of these schools. Indeed, an advocate of one school of thought will be prepared to refute all the other positions. A monist must try to explain why the two types of substances appear so different if they are fundamentally the same. If materialism is correct, why do thoughts appear so different from apples, stars, or even human bodies? If the idealist is right, an apple is but a thought in someone's mind. But if that is the case, would there be no apples if there were no minds? Bishop Berkeley offers the ingenious solution that the apple would still exist, because it is a thought in God's mind. This answer is one of the most original ones in the entire history of the problem, but many have found it difficult to imagine what a thought in God's mind would be. Finally the neutral monist runs into attacks from fellow monists who accuse him of being a dualist in disguise, and from dualists for being a monist.

A dualist has no difficulty explaining the differences in mind and matter. But if they are entirely different, how can they interact? How can we have a thought of an apple, and conversely, how can our thought influence us to pick the apple up? It is easy to say that there is a causal interaction, but how is this possible between things that are so different? As to the synchronized-clocks solution, would anyone really be prepared to accept the fantastic claim that these two worlds are entirely separate and just happen to be perfectly synchronized? (My apologies to Leibnitz and his followers who do accept this.)

Before the reader gives up in hopeless confusion, let us agree to set aside these age-old disputes. What concerns us is the problem of what Science can tell us of the nature of minds and of the relation between mind and matter.

THE SCIENTIST'S ATTITUDE

The average practicing scientist is a materialist.

We must understand this in the sense explained above, not in the entirely different sense of placing the value of material goods above all else, of which most scientists are not guilty. Rather, we mean that they are monists, considering the mental to be an offshoot of the material world.

Being a monist seems most natural as a working hypothesis. Why should we believe in a fundamental distinction, when we are not forced to? So many fundamental distinctions have turned out to be illusory, as for example the division between matter and energy. One could even argue that believing in this division from the outset is a violation of the rule of simplicity. Let us first try to explain all phenomena without admitting a split, and adopt the split only if the first approach fails.

If it is once granted that the scientist has a right, or perhaps a duty, to be a monist, it is clear that he will tend to be a materialist. It is in the world of matter that his greatest success has come, so he will try to explain minds as certain highly complex phenomena occurring in the presence of the most highly organized matter.

Let us interrupt this train of thought to compare the present problem to the previous one, that of vitalism. There too, the problem was whether or not we ought to admit a fundamental division. Again the dominant scientific attitude is mechanistic, against the division. We might even combine the two problems. It is fair to say that a dualist should be a vitalist. If you admit that mind is somehow drastically different from matter, you should hold that living things differ from inanimate objects. But you may be a vitalist without being a dualist; you may try to explain the behavior of minds in terms of living bodies. So, first of all, there is the materialistic scientist who holds that there is one unified subject matter for science, not admitting any sharp cleavage. Secondly, there is the vitalist who will insist on Life being fundamentally different. Thirdly, there is the dualist who will, in addition, demand that mental phenomena differ even from other living phenomena. The three positions are characterized by admitting zero, one, or two divisions.

To return now to the scientific approach to the problem of minds, we note that the materialistic position must look at the brain as a very complex machine and hold that any sufficiently complex machine is able to think. Let us investigate this hypothesis.

THINKING MACHINES

Perhaps the most significant factor in the recent developments of Man is the appearance of the first good thinking machines. These machines were designed for the purpose of carrying out complex computations, but they have reached a stage where they are much more than mere calculating machines.

Our giant mechanical brains are, among other things, excellent computers. They will carry out 4000 multiplications a second, to very high accuracy, with an excellent chance that no mistake will be made during this string of difficult operations. We can realize the enormity of the revolution when we compare this speed with that of other modern devices. Not so long ago the scientist marveled at the possibilities of the electric desk computer, and gave thanks for this wonderful labor-saving device. Today an electronic brain can achieve in a month what would take the human computer thousands of years with the aid of an electric desk computer. Problems that we would not have dared attempt twenty years ago are now routine tasks.

The progress in these electronic brains promises to be so rapid that the data here given will be sadly out of date by the time this book appears. There is no doubt that these servants of Man will replace him in thousands of tasks ranging from university research, through use by businesses and the Armed Forces, to the prediction of the weather and elections.

In spite of the fact that these machines excel as computers, it is misleading to refer to them as mere calculators. Past machines have been constructed to carry out specific tasks as efficiently as the technology of the day would permit. The present crop of machines is the first to qualify as "general-purpose" machines. This means that they can carry out any task for which the human operator can give precise instructions. If an intelligent assistant can, by following our instructions accurately, arrive at the desired goal, then so can the thinking machine, only much faster and with a negligible chance of error.

We have only begun to realize the potential of the machines already built. We use them to scan immense amounts of data, and to draw conclusions from them. They are used to supervise the work of human beings and to call various errors to our attention. They can be used as intelligent guessers, as when they were called upon to translate The Dead Sea Scrolls and other ancient documents, and to fill in words that have become illegible in time. When the process of making up instructions for machines became too hard for human operators, the machines were put to work to help make up their own instructions. And they are being readied for thousands of tasks that surpass the fondest hopes of their designers.

It is hard to deny that many of these tasks would be called thinking, and thinking of a high level, when carried out by human beings. If we heard that a certain person who is now playing mediocre chess is rapidly teaching himself better chess, or that he is learning how to prove mathematical theorems, or to manipulate complex strings of arguments, we would compliment him on his mental capacity. When we learn that this "person" was made by Man, and is just a mass of wires, electronic tubes, etc., it is hard for us to reverse our first opinion. If machines can carry out all these tasks that we associate with human thinking, are we ourselves more than mere machines?

The easy escape from being called "mere machines" is to say that we are conscious and machines are not. No doubt this carries considerable conviction, but it is hard to give a completely rational justification for this assumption. Suppose we met an alien species from another world, how would we decide whether they were capable of conscious thought? We would judge them by their acts. Suppose that a machine were built that could pass all our tests, would we still judge it to be unconscious? Science-fiction writers have made strong cases for the possibility of passing off a robot as a conscious being. As long as the machine is built in a shape that agrees with our idea of what superior beings should look like, we are likely to accept the being on the basis of its actions. When robots are "unmasked" in science-fiction tales, it is usually due to the robot being too perfect. But this is easily overcome. It is not hard to manufacture machines that will imitate human weaknesses as well as human strength. Indeed, we have already seen man-made creatures that give every evidence of having a nervous breakdown.

If there is to be any rational basis for our superiority to machines, it must rest on our being able to carry out certain types of actions that machines are incapable of. For the time being, at least, there is ample room for such a claim.

When we inspect one of the present mechanical brains we are overwhelmed by its size and its apparent complexity. But this is a somewhat misleading first impression. None of these machines compare with the human brain in complexity or in efficiency. While a computer will have tens of thousands of parts, the human brain has billions of cells. And no electronic engineer has yet begun to match the ingenuity with which these cells are "hooked up." It is true that we cannot match the speed or reliability of the computer in multiplying two ten-digit numbers, but, after all, that is its primary purpose, not ours. There are many tasks that we carry out as a matter of course that we would have no idea how to mechanize.

We have had to concede that any task in which we proceed according to strict rules, step by step, is a task better suited for our artificial brains. But this leaves the whole field of intuition, insight, and inspired guessing. An unfortunate by-product of the rise of Science has been that we tend to place small value on these "irrational" activities. The irony of this is that such irrational activity lies at the heart of Science. The entire activity of induction, and hence the formation of scientific theories, comes under this heading.

Human beings have the ability of solving problems by very

efficient short cuts, whose very nature they cannot explain. There is the mathematician who will see two similar examples from which he guesses a far-reaching theory, while at other times he will reject a hundred cases as mere coincidence. There is the doctor whose diagnosis of a rare disease is determined by processes he would never dare to explain to a group of medical students, or the mother who "instinctively" senses danger for a child. While many such instances are attributable to mere chance, some of them are no doubt due to a reasoning process too subtle to be described or analyzed.

Let us select the process of learning as one to illustrate this possible difference between men and machines. The crudest activity classified as learning is memorization. We admire men for their "large store of knowledge," which often means that they are a walking almanac, and we give them thousands of dollars in television prizes. Then we are right in having the deepest admiration for our electronic gadgets, since they can memorize the average book in an hour, and will be able to cite any passage at any future time with "lightning speed." It is a pity that mechanical brains are not permitted to compete in quiz programs.

But most of us would admit that learning facts is inferior to learning how to do things. Even in this area we find vast differences in learning. The simplest method is trial and error. We have a definite goal, which we recognize when we reach it, and we keep trying various approaches until one gets us there. A great deal of school learning consists in acquiring mankind's past experience in various trials and errors. But this is an activity in which machines excel. There are ingenious experiments under way in which machines will learn to carry out many tasks, even playing good chess, by such a learning procedure. So we cannot deny that machines are capable of a fair level of learning.

We have now put the machine on par not only with the quiz contestant, but also with the average college student. But an able student will leave his colleagues, and the present vintage machines, far behind. He acquires a mysterious "insight" that enables him to learn by a much faster and less wasteful method. It does not always lead to success, but the success is occasionally spectacular:

he will take one look at a problem and promptly guess the answer.

Whether this type of inspired guessing is teachable to machines is an open question. If it is teachable, then we will have machines that will make our present monsters look like small children. If the machine could combine our ability for remarkable short cuts in reasoning with its present ability for carrying out details and checking proposed solutions, it would far outdistance human reasoning in all fields. Whether this stage will ever be reached has to be left as an open question.

There is one more line of argument that we must examine. There are certain fascinating results of modern mathematics, proved in the 1930's by the Austrian mathematician, K. Gödel, and extended by the American mathematician, A. Church, which have considerable philosophical significance. A version of these results, for computing machines, was given by the English mathematician, A. M. Turing, which may be paraphrased as follows: We can give a general definition of what a machine is. This definition is so general that it certainly includes anything we have thought of so far, and it is difficult to see how we could build a machine violating this description. Given such a machine, we can always find a problem of the kind it can understand, which it cannot solve. Of course, there will be a better machine which can solve this particular problem, but we can ask it an even tougher question which it in turn cannot answer. These results prove rigorously that every conceivable thinking machine has its clearly defined limitations. And yet it is not clear that a human being, given enough time, couldn't solve all these problems.

These are the facts. What philosophical conclusion are we to draw? One possibility is that we really are essentially different from machines. We certainly appear to be different from all machines satisfying the definition. But aren't there undreamed of machines which are beyond the given description? If there are, we do not know what they could be. It would even be reasonable to say that if there are such, then they are in some sense alive. If we could build machines which overcome the barrier just mentioned, we may have to conclude that we have artificially created intelligent life.

Instead of arguing about this, it is much more fruitful to call anything that is man-made a machine, and then ask the same question, namely: Are we machines? Then what we ask is whether we can artificially build an intelligent being. The previous argument at least shows that this robot may have to be constructed in a manner different from anything we have dreamed of so far, that the gap is tremendous in our knowledge, in spite of these ingenious "thinking" toys. The final answer will have to be left to the Science of tomorrow, or perhaps of the next million years.

FREE WILL

Any treatment of the mind-body problem which ignored the problem of free will would be severely criticized. Are we really free to make decisions, or is this an illusion?

The famous arguments that there is no free will remind me of an old anecdote. A country hick takes his son to the city zoo. His son is full of curiosity and keeps asking his father the names of these strange animals. The father has a difficult time, because he can't read, and so is forced to keep guessing. Just about the time that the parent's imagination runs out his son spots a giraffe. "Dad, what kind of animal is that?" The father thinks, and thinks, and stares in more and more amazement, until he exclaims: "Son, there ain't no such animal!"

In the last analysis no amount of ingenious, involved, and incomprehensible argumentation will convince any of us that we have no choice as to whether we use our right or left hand to scratch our ear. It is a fact that we do make free decisions daily, and the role of Science is to explain this fact, not to explain it away.

Then what is the problem? One argument is that our will cannot be free, since the universe is determined. If the future is already determined, how can our supposed choice affect it? This sort of determinism occupied us in Chapter 11, and we decided that it is a pseudo-problem which arose out of a misunderstanding of the nature of scientific laws. We could restate the argument by saying that we cannot have a free choice because the Law of Nature says what the outcome of our choice will be. If it is already "written," then we have no real choice. The Law is not something binding, but a simple description of all events, past, present, and future. Among other things it describes how we choose. This is the only reason why our decision must be in accordance with it. It would be just as correct, and perhaps less misleading, to say that the Law of Nature depends on our choice, instead of the reverse.

In Chapter 11 we saw that this pseudo-problem of determinism is closely related to the very real problem of predictability, and this is the more reasonable version of the free-will problem. It is generally supposed that if a scientist knew enough about a person, he could predict what the person would do under given circumstances. If an outsider can do this, how can we say that we had a real choice? How can we be free to choose, if the scientist will tell us ahead of time what we will choose?

First of all, it is highly questionable whether human actions are fully predictable. But this problem we will postpone until the following section. Let us suppose for the sake of argument that our actions are predictable. If the scientist making the predictions could simply watch us from the outside and yet tell us what we are going to do, this would be surprising indeed. But this is not what is maintained. What *is* maintained is that if the scientist knows all about our physical make-up and past history, that is if he knows all internal and external factors influencing our decision, he will be able to predict the outcome.

As we have seen (Chapter 9), in predictions all relevant data must be known, besides having the necessary theories. So even if the materialists are right in supposing that mental processes can be reduced to nerve actions, and even if these theories were completely known, he would still have to know the initial data—external and internal. But then he would know all the facts that we know and would know the machinery we use for making decisions (remember we are assuming the materialist thesis for the moment, so that knowing our brain means knowing our mind), and then it is not at all surprising that he can predict.

I know my best friend very well, and it is entirely possible that I can guess his decisions with good accuracy. Does this mean that he had no free choice? For this I did not even need to know all the relevant facts, nor know any far-reaching theories; all I need are some good guesses based on past experience. After all, what is a scientific prediction but a good guess based on past experience? Looked at from this point of view the argument reads "You are not free to make a decision, because I can frequently guess what you are going to decide," which is ridiculous.

In short we do have free will. The only problem is to find out the mechanism by which we make decisions. This is a difficult and important problem for the psychologist, but not for the philosopher.

ARE HUMAN ACTIONS PREDICTABLE?

Let us turn to this very interesting question. The free-will dispute disturbs the average college student for two reasons. First, he is afraid that someone will prove his freedom an illusion; and secondly, he feels that predicting his actions is an inexcusable disturbance of his intellectual privacy. The former we have seen to be without point, but the latter is a real "danger."

Instead of asking flatly whether our actions are predictable, let us rather consider to what extent they are predictable. Certainly some actions can be predicted. Put a steak in front of a starving man, and you will have no trouble predicting his actions. On the other hand, we certainly cannot predict all actions, at least not for the time being.

There are two fundamentally different ways of trying to form theories about the actions of a given man. One is the behaviorist approach of looking at him from the outside only and forming generalizations about the pattern of his behavior. This is certainly the easier approach, and more fruitful in the short run. The second approach is to try to find out the internal factors as well, partly by studying his body and partly by questioning him. The latter approach (or approaches, if you care to distinguish the neurological and psychoanalytic approach to a man's interior) is much more time-consuming, leads to greater difficulties in forming theories, but is generally admitted to be the only hope for getting a complete theory.

As we have seen (Chapter 11), there are three problems in predicting: getting all relevant facts, forming a complete theory, and carrying out the calculations (deductions) necessary for prediction. We have also seen that all of these run into serious difficulties even before we come to human thinking. Here the difficulties are much greater. It is practically impossible to get thorough knowledge of the internal structure of a man without killing him. He is such a complex machine (if he *is* a machine!) that our theories will have to be very complicated, and hence will take a very long time to discover. Finally, the problems of calculating results of his decisions are far beyond anything we have faced so far.

Some types of human actions are more easily predictable than others. There are cases where few facts are relevant and simple theories suffice. These are the actions classed roughly under the heading of reflexes or habits. We start with these, and as psychology progresses, we can predict more and more. There are also differences between people in this respect. We say that there are "unpredictable" people. These are the ones whose habits form a pattern too complex for us at the moment. There is no doubt that considerable progress will be made in this direction, but when the complexity of the problem is fully understood, there seems little chance of the psychologist ever being able to interfere seriously with our privacy of thinking.

To summarize: Complete predictability of events seems out of the question even in the inanimate world. Prediction of human actions is perhaps the most complex problem facing the scientist. In this field we can expect only the very slowest progress, and we must expect that many types of decisions will forever remain outside the scientist's ability to predict.

Most important of all, we must remember that we are human beings. The very fact that a scientist observes us may alter our actions. This is the greatest complicating factor for the psychologist, and it is also the simplest way to demonstrate the relation between predictions and free will. When we predict where a stone will fall, we may tell it just what it will do, and it will still do it. But tell a human being that he will use his right hand to scratch his ear, and you are likely to find that he uses it instead to punch you in the nose.

SUGGESTED READING

Complete references will be found in the Bibliography at the end of the book.

The mind-body problem.
Broad [1], Chapter III.
Ryle, Chapter I.
Laslett.
Feigl [3].

Free will.
*Hume, Section VIII.
Moore.
University of California Associates.

Thinking machines.
*Wiener, pp. 113-167.
Kemeny [1].

14

Science and Values

"Would you tell me, please, which way I ought to go from here?" (asked Alice.)

"That depends a good deal on where you want to get to," said the Cat.

IN OUR SURVEY of scientific problems and problems arising out of Science there is one major gap. No mention has been made so far of the social sciences. However, the entire problem of methodology of the social sciences is intimately connected with the problem of how Science is related to Ethics. Perhaps the fundamental question a philosopher of Science must answer about the social sciences is whether scientific problems in the social sciences can be divorced from questions of value. Hence, it will be necessary to consider questions of value first, before we return to the social sciences in the next chapter.

VALUE STATEMENTS

In Chapter 2 we classified meaningful statements and arrived at a classification having two categories. First, there were the statements of Logic and Mathematics which were analytic and a priori. These statements did not have any factual content and their truth or falsity could be established without reference to facts. In effect, they were simply statements that analyzed the meanings of words. In the second category were the synthetic a posteriori statements of Science which furnished us with factual information and whose truth or falsity was based on facts. We might at this stage ask how value statements fit into this dichotomy.

First, let us agree on what a value statement is. We find a bewildering variety of statements classified under this general heading and we will have to agree for our purposes to put value statements into a common form. Under the general heading of a value statement, or an ethical assertion, one will find such various items as "This action is good," "You must do so-and-so," "The action of this group is highly undesirable," etc. Any type of command, or assertion that something is good or desirable, or bad or undesirable, has a value connotation.

Let us agree to put any such statement in the form "such and such is good." A command instructs us to carry out a certain action. The content of this command may be interpreted as saying that this particular action is good. Similarly a statement of what is desirable or undesirable can be put into this standard form just indicated. Of course, many other standard forms could serve our purpose equally well, but this arbitrarily chosen standard form is as convenient as any.

Let us choose as our example "That Jim Jones studies Mathematics in college is good." If we omitted the final clause "is good" we would have a simple factual assertion and hence something that belonged in the second, or synthetic, category. Where does the entire statement belong? Surely it is not an analytic statement; the truth or falsity of this statement cannot be established purely by understanding the meaning of the words involved in it. On the other hand, we would be hard pressed to decide what type of factual information is relevant to this statement. We may observe Jim Jones, we may observe the college in question, and we may make a careful study of the nature of Mathematics; in short, we may have all relevant facts at our fingertips, and yet it may not be clear whether this action is good or bad. Somehow one feels that the type of decision necessary to judge the statement is entirely different from either the processes of logic or the processes of the scientific method. We are forced then to admit a third category—that of value statements. That these statements are somehow essentially different from scientific assertions can also be seen from the fact that they are often phrased in the form of a command. One of the very deep problems in Philosophy is to rule on the status of these value statements and to discuss the entire question as to how one can decide whether they are true or false. However this problem is certainly beyond the scope of our study.

Actually, we have good evidence that value statements were discussed by mankind long before the dawn of Science. The Ten Commandments are among the earliest in our western civilization, and eastern civilizations have much older records. It is very likely that as soon as human beings gathered together in social groups, problems of right and wrong were among the most frequently discussed issues. Indeed, if the most generally believed account of the creation of man is accepted, then our value statements did not exist from the beginning of the human race, but they followed this beginning only by a matter of days. We are told that the very first crisis in the history of mankind arose over the issue of whether Man is to know right from wrong. It is certainly true that pre-occupation with right or wrong is characteristic of the human race, and we are all called upon a thousand times each day to exercise the mysterious gift of making such decisions.

Since the question of the nature of these statements and the method by which decisions are made concerning them raises the whole theory of values, we will for our purposes have to trust our intuition on both these matters. We will restrict ourselves to the much more limited question of value statements occurring in Science, and what role, if any, scientists play in making decisions concerning them.

THE FACTUAL BASIS OF VALUE STATEMENTS

The major issue that we have to decide is whether value statements can be established by scientific methods. It is now quite generally accepted that this is impossible.

The scientific theory that has most often been used as a proposed source for value statements is the theory of evolution. Early admirers of this theory proclaimed an ethic of competition "derived" from the struggle for existence. More recently the same theory was used by critics of this ethical system, like Julian Huxley. Huxley claims to have derived most of the accepted ethical norms of Anglo-Saxon morals from the theory of evolution. For example, he "proves" that everyone ought to have an equal opportunity to develop himself. His general form of argument is to show that if a certain ethical maxim is obeyed, then evolution has a much better chance of developing rapidly. From this he concludes that the action is indeed desirable.

There are three types of arguments one can level against Huxley's position. First, there is a tacit assumption that the present form of the theory of evolution is correct. Indeed Julian Huxley himself criticizes the earlier attempts at deriving value statements on the grounds that they were based on an incorrect theory of evolution. Secondly, it is questionable whether Huxley's arguments really establish that these ethical maxims are the only ones that will expedite the theory of evolution. Thirdly, there is a hidden premise in all of these attempted derivations of value statements. There is a major gap between establishing that a certain action will expedite evolution and the conclusion that this action is desirable. The missing premise, of course, is that expediting the theory of evolution is good. Actually one could show that a stronger premise than that is necessary, namely, that expediting the theory of evolution is the only good or the most important good available to mankind. Without the stronger form of the premise, one might argue that, while an action that expedites the theory of evolution may be good, there may be other considerations that overrule this.

Let us now consider the same argument in the abstract. We have used the version of the simple rule—you cannot get something from nothing—in Chapter 2. In order to derive factual conclusions, one must have some factual content in the premises. Similarly, to derive conclusions on any one subject matter, there must be premises that deal with this subject matter. If we take a typical logical deduction, such as "All whales are mammals, and Moby Dick is a whale; therefore Moby Dick is a mammal," we note of course that both "Moby Dick" and "mammals" are mentioned in the premises; otherwise these could not have been connected in the conclusions. Similarly, if the conclusion is to contain a statement about something being good, then the word "good" or one of its synonyms must occur in the premises. We cannot derive a conclusion about questions of value by purely logical means unless the premises themselves furnish values.

Before we accept this conclusion we might try some argument against it. Consider the deduction "All eagles are birds and Mary Bell is an eagle; hence Mary Bell has wings." We could say here that a conclusion was drawn about wings when no premise talked about wings. However, this analysis of the situation is not correct. It is understood in the argument that birds are animals which, among other properties, have the attribute of having wings. Hence wings were mentioned indirectly in the first premise. One might hope to get around difficulties mentioned above by claiming that, although factual statements do not use the word "good" explicitly, they may make indirect reference to this. However, this argument is very weak. Factual statements are formulated in terms of empirical concepts and these do not label anything as good, either directly or indirectly.

There is still one more type of objection to the above argument. One can point out that the statements made previously are not, strictly speaking, correct. Suppose that we have the premise, "Peter is a tall man," and that we conclude from this that "Peter is either tall or handsome." This is a perfectly correct argument. Its conclusion refers to someone being handsome, even though there is nothing directly or indirectly mentioning this concept in the premises. Quite similarly, one could draw a conclusion that involves the word "good," from premises that are factual statements. Suppose that we modify a previous argument, as follows: "All whales are mammals and Moby Dick is a whale, therefore either Moby Dick is a mammal or the Ten Commandments should be obeyed." This ingenious argument, due to C. G. Hempel, establishes one point: General arguments about the form of statements and about the forms of deductions are very hard to formulate correctly. We showed that the type of glib proofs all of us are tempted to give are generally full of discrepancies. But it does not change the essence of the conclusion we arrived at earlier. Certainly all of us feel that the type of conclusion just drawn is not suitable as a foundation for ethics. One can still modify the previous argument to show that no categorical statement of the form "so and so is good" can be deduced from factual statements. The argument is now somewhat more intricate and more technical, but the essence of the conclusion can be maintained.

For practical purposes we can conclude that scientific statements in themselves cannot serve as a source of value judgments. It is interesting to note that some of the most distinguished philosophers of science have fallen prey to this so-called "naturalistic fallacy." One may suppose that the desire to find a firm foundation for ethics is so strong that it will blind even able thinkers to errors in their logic.

The opposite fallacy is now confronting us. We may be tempted to conclude that for the above given reasons Science is entirely irrelevant to Ethics. But we can quickly find examples that will convince us that we are mistaken in this point of view. Suppose that we have decided that building a bridge of a certain caliber is a highly desirable goal. We now try to decide what we ought to do. Surely at this point we will consult a group of able engineers; we will not ask the ethical philosopher to draw up the blueprints. Again let us suppose that we have decided that eradicating multiple sclerosis from the United States is a highly desirable goal. We do not at this stage call upon our moral teachers to tell us what every man should do, but rather we consult the biological scientists and physicians who are best qualified to give us advice. Here we find scientists in the role of telling us what we ought to do. Just how does this take place?

The essential point here is that we have already decided on a goal and we are searching for means. If we try to schematize the type of argument used, it may look somewhat as follows: "It is good that multiple sclerosis should be eradicated from the United States. The following is a list of actions that are necessary to eradicate multiple sclerosis from the United States. Therefore these actions should be undertaken." It may be said that the most difficult set of premises in here are furnished by the scientists. However the premises do contain one value statement—namely, the desirability of the eradication of a dread disease. What we know is that once a single ethical premise is given, Science may furnish us with a long list of imperatives as a guide for our everyday actions. Indeed I will try to maintain the position that Science plays an important role in Ethics; namely, once ends are specified, Science furnishes us with the means of achieving these ends. To put it in other words: if some list of ultimate goods has been specified, Science will give us a long list of inferior goods that are necessary to complete the scale of values.

The difficulty confronting us in the naturalistic fallacy reminds me of an anecdote which is attributed to the logician, Kurt Grelling. He told a story about a friend of his who found the secret of perpetual motion. He described in detail a highly complex machine which made use of all the latest engineering techniques. After an elaborate description of this marvelous piece of gadgetry, he confessed that as a matter of fact the perpetual motion machine was not entirely perfect. It seems that out of this great machine there sticks one small lever. The only step that is yet to be carried out in the construction of the perpetual motion machine is to find a method by which this minute lever will wiggle continuously.

In the same way we feel that the tremendous and impressive structure of Science in itself is totally useless as a basis for ethical judgments. However, if we furnish it with this small wiggling lever, which in our case is a statement specifying our goals, we will derive tremendous value from the body of knowledge furnished by Science. The quotation due to Alice at the beginning of this chapter is highly relevant. If you ask a scientist just which way one ought to go, the only answer the scientist can give is that given by the Cat: "That depends a good deal on where you want to get to." We must decide by means other than scientific just where it is that we are heading. But once this decision is made, Science can tell us how to go there, in what direction to go, and the fastest and best means of getting there.

ENDS AND MEANS

Let us consider a very simple example where we can study the relationship between value statements and factual statements. We have been asked to design an airplane, and the problem that confronts us is the material to use for the body of the plane. Let us suppose for simplicity that we narrowed the choice down to five possible materials and combinations of materials. Thus our decision is to be a choice of one of these five.

One might suppose that some of these decisions can be made without attention to any value considerations. For example, we might say that if the body is made from a material that is too heavy, the plane will not fly. However, we have begged the question: Why should it fly? If we have decided that it is desirable for the plane to fly, we have already made a value decision. The naturalistic fallacy no doubt arose out of such cases where the value judgments to be utilized were so obvious that they were never consciously considered. We find ourselves in a situation where a decision is impossible without further directive. We must find out just what it is that this plane is to achieve.

There are conflicting ends that one may have in mind. We may try to have a workable plane at minimum cost, or a plane that can fly as high as possible, or one may aim at maximum speed, maximum durability, or maximum safety. It is entirely possible that these divergent goals would be best achieved by divergent means; hence the decision of which of the five materials to use will depend on the goal we have in mind.

The previous argument showed the need for value judgments in making our decision. It is, however, equally clear that the value judgment in itself does not suffice. Suppose that we know exactly what we want to achieve. Say that we want to achieve maximum speed without sacrificing a certain amount of safety and yet keep the plane's cost as low as the above goals permit. This may very well narrow the choice down to a single material, but no value consideration will tell us which. We have to consult a competent scientist to choose among the five materials to achieve this goal. Indeed, this is the way scientists work for the military. Some general, or group of generals, will decide what achievement they want from a plane, and then they consult scientists who will tell them whether a plane with these specifications is feasible, and, if so, how costly it would be and how long it would take to build it. Once they have agreed on feasibility and on a reasonable budget, the scientists have to decide what means will best achieve the aims, the ends that the generals have specified.

We will see later that this division into means and ends is somewhat oversimplified, but it will serve for our present purposes. We know that one of the fundamental roles of Science is to construct causal laws. These have the property that if one knows the present situation completely, the law will help us predict the future. In the type of situation that we are now considering, the present is not yet completely determined. It still hinges on a decision we are to make. All we can expect of our causal law is to tell us what the future will be like, depending on the decision we now make. For example, in the above illustration, a causal law will tell us for each of the five possible decisions what the outcome will be. Then our choice boils down to deciding which of the five possible outcomes is best, and this is a value judgment.

In a general situation the connection between value statements and Science is quite similar. We are in the present which will be determined only after we have made a decision or a series of decisions. Science can tell us two things: First, it can tell us what decisions are open to us; and, secondly, it can tell us what the future will be like, depending on the way we make our decisions. Then we may think of value judgments as choosing between possible future states. Without value judgment this choice is impossible. Science can only tell us that such and such states can be achieved; it cannot distinguish between desirable and undesirable states. On the other hand, value judgments in themselves are not enough. They can tell us what states to choose, once we know what states are feasible, but they cannot tell us what the feasible states are, nor can they tell us what action in the present will achieve the desirable state in the future.

We may summarize the previous discussion by saying that value judgments enable us to choose among ends. On the other hand,

Science tells us which ends are feasible and how we can achieve these various ends. In other words, Science provides us with a list of feasible ends and with the means of achieving these.

We can draw an analogy between the role played by Logic or Mathematics in Science and the role Science plays for Ethics. Mathematics alone cannot establish a scientific theory; but, once a theory is given, it can deduce a large number of predictions from these. In other words, although Mathematics alone cannot give us factual truths, it can give us factually true statements from other factually true statements. Similarly, Science alone cannot give us value judgments. However, once the ends are furnished for us, Science can deduce the necessary means. Hence, once some value judgments are available, it can give us additional ones.

A COMPLETE SCALE OF VALUES

In an ideal situation mankind would arrive at a complete scale of values through ethical considerations. This scale is to be complete in the sense that, given any conceivable set of ends, the scale will pick out a unique one as most desirable. Also, in this ideal situation mankind would have an omniscient Science according to which it would always know what ends are feasible and what the right means are to these ends. Then, at any stage in the history of mankind, Science would provide Ethics with a list of feasible ends, Ethics would choose among these, and then Science would proceed to provide us with the necessary means. Let us now examine how far short we fall of this in our everyday proceedings.

First of all, we have noted the shortcomings of Science. It is supposed to fulfill the double role of providing Ethics with a complete set of possible ends, and it is supposed to be able to find the proper means for these ends. However, Science—as actually available to us—must fall short of both of these aims. We know that Science is incomplete and hence cannot give us any such overall knowledge; and we know that, even if the theories were complete, our means of logical deduction are so limited that the kind of universal question here asked is beyond our present means of mathematical deduction. However, these difficulties have been discussed sufficiently in previous chapters. Let us now assume, contrary to fact, that we have a perfect Science available to us and let us explore the difficulties that still confront Ethics.

We may fruitfully carry out our discussion in terms of a well-known ethical system. The system of philosopher J. S. Mill, "Utilitarianism," is perhaps the most frequently discussed ethical system in our culture. It identifies "good" with the production of the greatest possible happiness for the greatest number of people. In comparing two courses of action, the one that will produce more happiness for more people is better. If I am fortunate enough to inherit a million dollars overnight, I am confronted with the problem of what to do with it. According to utilitarian principles I must not keep the sum entirely to myself, for this will cause one person a great deal of happiness but no other person will be made happy by it. On the other hand, dividing it among a million people, giving each one one dollar, would also not be considered very good—while a great many people are affected, the amount of happiness given to them is negligible. Presumably a good utilitarian would arrive at some sort of compromise, where a reasonably large number of people would each receive a fairly large sum. The precise distribution would have to be determined by Science on the basis of psychological and sociological studies of the amount of happiness people gain as a result of gifts of various sizes.

The obvious criticism of this ethical system is that it requires a numerical measure of amounts of happiness. But this is not an ethical problem, and we will ignore it here. We will suppose that some highly reliable quantitative measure of the happiness of individual people has been found and that this measure has the desirable property that the amount of happiness of one person can be compared with the amount of happiness of other people. It would then appear that we have a workable, complete scale of values. This, however, is a mistake.

Let us return to the example of the million dollars that is to be divided among a group of people. Suppose that we consider dividing it evenly among a million people, a thousand people, or keeping it all to ourself. We may further suppose that, according to our happiness scale, this would mean providing one unit of happiness to a million people, or a hundred units of happiness to a thousand people, or ten thousand units of happiness to a single person. One may be tempted to conclude that the first method of division is best, since it creates a million units of happiness, while the others create only a hundred thousand, or ten thousand units. However, this is *not* what utilitarianism tells us to do. It does not say that we should maximize the total amount of happiness, but that we should try to maximize the number of people we make happy and the amount of happiness provided to these. These two goals, however, are inconsistent. The more people we try to make happy with the million dollars, the less happiness we can provide to each one of them.

As another example we will consider the dilemma confronting a young man in choosing his college. There is a school near his home where his parents would like him to go; on the other hand, his best friend is going to an out-of-state school and would like to have him accompany him. As a compromise he might go to a school that is fairly near and one to which his friend would also be willing to go. By going to the nearest school, he would make his parents happy; on the other hand, by going to the school farthest away, he would make his best friend extremely happy; and there is the compromise solution where everyone would be satisfied, but no one would be very happy. This dilemma is clear cut. Should he make one person very happy, two people fairly happy, or satisfy all three people without pleasing any one too well? It is clear that utilitarianism cannot help us out of this dilemma.

A complete scale of values is intended to provide a simple ordering of the states of the world. When Science provides us with a list of possible future states, the scale of values must enable us to choose among these. Utilitarianism, as proposed, has a scale of values, but in most cases it does not enable us to make a unique choice. It provides us with a partial ordering of the states of the world, but not a simple ordering. If one action applies to more people and makes each one happier than a second action, then it is clearly preferable. But if the former makes more people happy, but makes each person less happy than the latter, then utilitarianism does not lead to a decision. Of course, the most general situation is vastly more complicated. Suppose that the first action makes ten people happy, distributing units of happiness ranging from one to ten-thousand; and the second action makes twenty people happy distributing units of happiness ranging from ten to twenty—how are we to decide which action is better?

The next question that presents itself is whether utilitarianism could be corrected so that it does become a complete scale of values. The answer is that this can be done easily, in many different ways, and it is entirely unclear which of these ways the authors of utilitarianism have in mind. For example, the rule that tells us to maximize the total amount of happiness is a rule that will lead to a simple ordering. In the case of the million dollars, it will tell us to distribute it among a million people. In the case of the student choosing among colleges we would have to find out the units of happiness assigned by each action. If going to the school of his friend's choice would give his friend a hundred units of happiness, going to the home-state school would give each of his parents thirty units of happiness, and going to the compromise school would give each person fifteen units of happiness—then the first alternative is preferable.

Another possible rule could aim at as even a distribution of happiness as possible. This might tell us to examine all the people concerned and pick out the person getting the least amount of happiness and try to maximize this. This rule would again tell us to distribute the million dollars evenly among a million people, but would lead to a different choice in the college case. Going to the furthest school would give no happiness to the parents, while going to the nearest one would leave one's best friend out of the picture—hence the correct choice now is to go to the compromise school. As a matter of fact, this type of rule would often lead to a compromise.

Still another rule might tell us to take the average per-capita happiness, to square it, and to multiply it by the number of people involved. In this case we would be told to keep the million dollars and to go to the school which our best friend chose. In general, this rule would tend to make a few people very happy.

There is no end to the ways in which utilitarianism could be corrected to become a scale of values. However, any one of these proposed definitions has very serious objections to it from an ethical rather than a logical point of view. Utilitarianism sounds very plausible as long as it is incomplete. Once we complete it, it loses its ethical appeal. We find that, quite aside from the incompleteness of Science and of our means of deduction, there are fundamental difficulties in forming a complete scale of values. It seems as difficult for us to agree on any such complete scale as it is difficult to find a complete set of scientific theories.

It is also important to note that in this argument we have demonstrated several times the relevance of Logic, Mathematics, and Science to building a complete scale of values. A complete scale of values is one that enables us to give a simple ordering of states of the world. We must call upon Science to tell us what the possible states of the world are and to help us in measuring the various factors that seem relevant to us. On the other hand, we must call upon Logic and Mathematics to assure us that our scale of values is consistent and that it really leads to a simple ordering. Just as Logic and Mathematics played an important role in the development of the scientific method, so Logic and Mathematics, together with Science, serve as important servants in the study of Ethics.

SUGGESTED READING

Complete references will be found in the Bibliography at the end of the book.

Value statements.
- Broad [3].
- *Ayer.

Relation of Science and Ethics.
- Einstein.
- Frankena.
- Huxley.
- Broad [2].
- Hull.
- *Stevenson, Chapter XV.

15

The Social Sciences

"Speak English!" said the Eaglet. "I don't know the meaning of half those long words, and, what's more, I don't believe you do either!"

THE BASIC ISSUE in the philosophy of the social sciences is the question of whether Man can be studied by the same methods that apply to lower beings or inanimate nature. This is perhaps the most debated question in the philosophy of Science and we must consider this as our last major problem.

THE STATUS OF THE SOCIAL SCIENCES

It is certainly true that the physical sciences have developed to a stage far beyond that of the social sciences of today and even of the anticipated future. If one examines the laws of the physical sciences and compares those with known laws in the social sciences, they seem to belong to a different species. A typical law in the physical sciences is stated precisely, usually in mathematical terms, and is quite free of ambiguity. It has been tested repeatedly and has withstood the tests. The usual law in the social sciences, on the other hand, is ordinarily couched in Big Words and a great deal of ambiguity. The law is usually presented with many qualifications and excuses. There are probably several exceptions known to the law, but the law still has its advocates as the best we know so far. The law in the physical sciences has enabled us to deduce precisely certain predictions which have been verified.

While predictions have been attributed to the social law, the chances are that they simply reflect prejudices or commonsense knowledge of the authors of the law, and that these predictions have not been deduced from the law itself.

If we question the leaders of the field, it is quite likely that they will give us many reasons why the social sciences are so far behind the physical sciences. They will point out that the physical sciences deal with such simple objects as an atom, whereas the social sciences have to deal with human beings, singly or in groups. It will be pointed out that the basic tool of the physical scientist is the laboratory experiment, but such experiments are not permissible in the social sciences. Such reasons are legitimate methodological arguments. They may convince us that it is much harder to find laws in the social sciences than in the physical sciences.

In addition to methodological arguments we will also find frequent discussion of why, in principle, it is impossible for the social sciences to become truly scientific. We are told that inanimate objects obey laws, but human beings do not. We are told that within animate nature prediction is possible; with human beings, due to their free will, prediction is, in principle, impossible. Above all, we are told that social sciences must be basically different from physical sciences, since questions of value enter the former but not the latter. For these reasons, a purely scientific approach to the social sciences is supposed to be impossible.

However, we have already examined arguments of these sorts and found them untenable. The first argument is based on the mistaken belief that nature "obeys" laws in the sense in which human beings obey them. Due to this mistaken belief it is natural to suppose that human beings could disobey laws of nature in the same way that they disobey human laws. However, as soon as we realize that the laws of nature simply describe what actually takes place, this type of objection becomes absurd. Why should it be harder in principle to describe the actions of human beings than the actions of atoms? If anything, it should be easier, since we have direct experience of what it is like to carry out human acts. The second argument, concerning predictability, was considered in Chapter 13. We found that free will is in no way incompatible with the possibility of prediction. Indeed, there have already been many fine examples of predictions in the social sciences. Yet none of us feel that our free will has been infringed upon.

The final argument, concerning value statements, is at the heart of a great many of these objections. This problem was considered in Chapter 14. It is, of course, entirely possible to insist that the social sciences should deal with questions of value. If the purpose of this activity is to find answers to value questions, then it is certainly true that the social sciences will have to be fundamentally different from the physical sciences. However, this is a verbal argument. It cannot be denied that it is in principle feasible and in practice important to find scientific laws which will enable us to correlate human means to human ends. This type of activity is appropriately described as social science, and it can be handled by the scientific method discussed in this book. It would seem to me most fruitful that only this activity should be called "social science," since the advocates of the other type of undertaking admit that it is not manageable by scientific methods. Of course, we run up against vested interests. There are many members of social science departments in our universities who teach what is right and what is wrong, and they would be afraid that, if these activities were labeled as unscientific, they might lose their positions.

Closely related to this debate is the question as to whether social sciences should deal primarily with historical laws. In this they would serve primarily to relate what has already taken place, to classify the various acts, and to try to rationalize the motives that led human beings to act in certain ways. In any science the gathering and classification of facts is an important first step. But if we can learn any lesson at all from the physical sciences, we must realize that this is a bare beginning. Real success comes only after this stage has been left for the higher activity of theory formation.

We have noted earlier that it is questionable whether classification of facts can ever take place without having some theory in mind. Indeed, historians usually have a thesis that they would like to establish, and they present historical facts to serve as evidence for their proposed hypothesis. In a field that is very unprecise, this is a dangerous kind of activity. The temptation is great to twist the historical facts to fit a proposed hypothesis. Indeed, we find social scientists constantly accusing each other of having done precisely this.

Let us, therefore, conclude that in principle there is no difficulty in applying the scientific method to the social sciences, but in practice we run up against severe difficulties. Let us now turn to the consideration of these difficulties.

METHOD IN THE SOCIAL SCIENCES

We have analyzed the scientific method into three major stages: the formation of theories, the deduction of consequences, and the verification of predictions. We will consider how these activities take place in the social sciences, taking the activities in reverse order.

First, there are obvious difficulties confronting us in verifying a given prediction. Let us suppose that a given theory predicts that, when ten million people are isolated on a desert island for a hundred years, certain social phenomena will take place. Clearly there is no practical way to verify this prediction. One may have to wait for a historical accident for such an event to take place, and the waiting period may be forever. The social scientist is quite right in pointing out that the physicist has a great advantage with his controlled laboratory conditions and his ability to observe under ideal conditions.

Laboratories for the social sciences are now in existence at many universities throughout the world. However, it is a difficult question whether human beings isolated in a specially built observation room behave the way they would normally behave. The experimentalists among the social scientists have been accused of searching for laws that will apply only in these isolation booths. Nevertheless, it is reasonable to believe that, for certain simple types of actions, these experiments might be entirely satisfactory. The history of Science points out that each branch must start with the formation of simple laws which are later extended to cover the really interesting cases. But even among the experimentalists one finds that it is hard to reproduce an experiment exactly.

Take a specific example. A dispute has arisen among certain experimental psychologists as to how subjects behave under given conditions. An interesting experiment which is described by W. K. Estes, one of the pioneers in psychological learning theory, is the following: A subject is asked to guess whether a light in front of him will go on within the next five seconds. After he has guessed, the light may or may not be turned on, according to a predetermined pattern. Suppose that the experimenter turns on the light at random, with the light coming on 75 per cent of the time. If the subject would know this fact, and if he maximized his chances of guessing correctly on any one try, he would guess "Yes" each time. However, this is not what the subjects actually do. It was shown that, in the long run, the subject will guess "Yes" 75 per cent of the time and "No" the rest of the time. As a result of this, he will be right only 62.5 per cent of the time, thus doing considerably worse than is possible for him to do.

Other experimenters have carried out "the same" experiment and have found that the subjects eventually guess only "Yes." However, when we look into the experimental procedures, we notice certain differences. In the first group of experiments, the subject is asked to go on guessing and is not allowed time to consider the best way to proceed. In the second group of experiments, the subject is asked to pause after a certain number of trials and is given a chance to consider whether what he is doing is the best that he can possibly do. The real question now is whether the two experimenters have carried out the same experiment or not. I would certainly feel that these two sets of experiments are drastically different and that the results show reactions to different conditions; but the fact that eminent psychologists disagree on this point exemplifies a basic source of trouble in experimental social science.

Last but not least is the difficulty of trying to verify a prediction about the future. The fact that you are carrying out this verification may alter the actions of your subject, either to bring about verification or to destroy it.

The difficulties that have been given least thought are the difficulties in deducing consequences of laws in the social sciences. Of course, it has been pointed out frequently that precise deductions from vague laws are impossible. It is not worth dwelling on this question. Let us consider only those rare laws in the social sciences which have been precisely stated. We know that deduction in any science amounts to the solution of simple or complex mathematical problems. Since the social sciences are still in their baby shoes, it has been assumed that the solution of whatever mathematical problems may arise in the social sciences must be elementary. This has been the principal reason why most of the leading mathematicians have ignored applications to the social sciences. However, this is far from being the true picture.

To understand this assertion we must consider the nature of mathematical knowledge. In any actual application one deals only with a finite number of objects. Let us classify such applications typically into three categories: In the first category one deals with a small number of objects, let us say five. In the second category one may have 5000 or 50,000 objects. In the third category one would typically have five billion or five billion billion objects. The first type of problem is the kind which can be solved by elementary mathematical methods, by working through all possibilities one at a time. It might seem that the hardest type of question is that dealing with billions or billions of billions of objects; however, here is where the calculus has made a tremendous contribution. We have found that for many such problems one may assume that there are infinitely many objects, and hence apply powerful mathematical tools developed just to answer such questions. For example, in physics, when we wish to measure the velocity of an object, we assume that, at least in principle, we can make infinitely many observations of the position of the object. Hence the methods of the calculus are applicable.

It is the intermediate region of five thousand to fifty thousand objects where neither of these methods is applicable. The numbers involved are too large to try all possibilities, and yet the numbers are too small for the analytic methods of the calculus to lead to accurate results. As a matter of fact, the only reasonable description of this mathematical situation is that we have not yet developed the kind of mathematics that is needed in the intermediate region. Because of this, problems in the social sciences are not easier than in the physical sciences, but actually considerably harder. Here we have a good example of progress in a branch of science having to await mathematical progress. Unless more of the really able mathematicians take an interest in these problems, this field will be very slow in developing.

Perhaps it is misleading to say that there are few really interesting and precise laws known in the social sciences. It may be quite unfair to the social scientist to say that he has developed only the most trivial theories. A fairer statement might be that only some of the more trivial theories have been treated to a degree where one can judge their adequacy. Many more laws have been proposed, but one very quickly runs into mathematical problems that exceed our ability for solving them.

Finally, we have the problem of theory formation in the social sciences. Here we run into the greatest difficulty facing the social scientist, and this is a handicap in motivation. The physical sciences developed at a time when no one had any clear idea of the potential of Science. As a result, physicists were content in finding laws describing extremely simple and basically uninteresting phenomena. But since then the physical sciences have shown us that scientific development can lead to wealth and power for nations. The social scientists are understandably impatient in reaching a status where they themselves can make an active contribution to the development of civilization. It certainly takes remarkable human patience and humility to study the behavior of five human adults solving a childishly simple problem, when the social scientist has confronting him the problem of predicting inflations or financial crashes or of making predictions as to how a given nation can increase its wealth and position in the world. Yet all precedent points in the direction that it is the former that will lead to progress, and not the latter.

It is also difficult to keep emotional overtones out of hypotheses. Any human being is anxious about the ethical implications of actions. It is hard to formulate an unemotional version of the prospective results of human undertakings. In the history of the laws of celestial motion we realize how fortunate the astronomers were in not really caring what the exact paths of celestial bodies were, as long as they could describe them precisely. As a matter of fact, when some of these questions acquired political or religious overtones, they led to bad science. When it became a heresy for Galileo to maintain a view contrary to the accepted doctrine, it certainly had the effect of stifling the progress of physics. Fortunately, in this case there were scientists in other countries who could proceed in freedom. The difficulty in the social sciences is that the pressure to arrive at "desirable" results comes not from the outside but from within the scientist. He himself would like to believe that certain kinds of actions will produce desired results. He cannot help but search for laws that will enable him to make these predictions.

The physical scientists also had the great advantage that no generally accepted terminology was available for the phenomena they were studying. In many cases they had to coin new terms, and it was a matter of little interest to their colleagues, and to the public as a whole, what words they used. The social scientists face the dilemma of either trying to describe well-known phenomena in technical terms, or of falling heir to all the vagueness and ambiguity of everyday language. There is some justice in accusing the scientist of coining a new word unnecessarily for something well known. But if this well-known act is tied up with prejudices, hopes, and fears, a technical term may turn out to be more suitable. Thus the social scientist must create his technical vocabulary not in a vacuum but in a space that has been filled by useless and dangerous words.

Above all, the social scientists are right in saying that the phenomena they study are basically more complex than the phenomena of the physical scientists. Because of this it is hard to see where to start in the formation of fruitful concepts. Although the distinction between inertial and gravitational mass is a subtlety to be mastered only by twentieth-century genius, it was clear from an early stage that somehow the weight of an object affected its motion. In the study of motion it must have been clear from the beginning that the speed with which an object moves was important, though it took centuries of progress to realize that acceleration was more basic than velocity. In the study of human beings it is difficult to find any clear-cut concept that is certain to play a basic role in the study of Man.

To summarize, we find that laws are harder to form because of the tradition of vagueness, ambiguity, and emotive overtones in the subject matter, and because of the inherent complexity of human beings. We find it difficult to make predictions, because we quickly run into mathematical problems that are too hard for us. Even if we are fortunate enough to make predictions, we are not in a position to carry out carefully controlled experiments. We may make predictions that will take a long time, if not forever, to verify. Even in the cases where verification is possible, the number of cases we can study will be of a much smaller order of magnitude than in the physical sciences. Considering all these factors, it is not surprising that the social sciences have developed much more slowly than the physical sciences.

AN EXAMPLE

To make the ideas here presented more concrete, let us consider an example of a theory in the social sciences. I will take this example from economics, choosing a model of an expanding economy due to the late, very noted mathematician, John von Neumann. This theory presents certain hypotheses as to how an expanding economy, such as the present United States economy, behaves in equilibrium.

The terms "expanding economy" and "in equilibrium" are used in a precise sense in the theory. They correspond reasonably well to the meanings of these words in everyday language, but there are also dangers in carrying everyday meaning too far. The economy is represented by certain production processes which enable it to produce various goods in varying quantities. The theory makes predictions as to the amounts of these goods produced, and as to the prices assigned to the goods. The basic theorem states that there is at least one way for the economy to expand in equilibrium. There are only a finite number, and usually a very small number, of different ways in which it can function in equilibrium. The largest possible rate of expansion is presumably the one in which it will function under ideal circumstances. There are other theorems of considerable interest to the economist, such as: in equilibrium the rate of interest in the economy equals the rate of expansion.

This theory has been attacked on many grounds. For example, it presupposes that there is a single interest rate in the country. One can certainly show many examples where various interest rates will be applicable to different segments of the economy. Of course there are many ways of getting out of this difficulty. The simplest is to say that our economy is not in equilibrium. Or one can say that the United States is highly complex, and actually it is several economies in one, each of which may be in a different equilibrium. These explanations are both plausible and potentially dangerous. After all, if we carry this far enough, no matter what example is brought up in attempting to disprove the theory, we could talk our way out of it. It is not too unreasonable to maintain that whatever the sense of "equilibrium" is, no economy is exactly in equilibrium. But if we take this position, then the question arises as to whether such theories are of any value.

Let us compare this with Newton's Law of Inertia, which states how a body behaves if there are no forces acting upon it. It is certainly true that such a situation can never exist in practice. No matter what body there is, there are at least gravitational forces acting upon it at all times. The most one can maintain is that a body far enough from the major masses in the universe will be approximately free of forces, and hence Newton's law should apply to it approximately. The same is true for any oversimplified theory, as von Neumann's no doubt is. One should have a clear-cut notion of when an economy is approximately in equilibrium, and then one could test whether von Neumann's predictions actually hold true. The danger in this is that one may have to wait forever for such an economy to exist.

Next let us consider the problem of prices in the economy. Under many circumstances the theory will give us a unique set of prices that must be applicable to each stage of the expansion.

Yet it is entirely possible that, as a matter of fact, none of these prices correspond to actual sums of money paid out by purchasers. For example, this theory, as well as many similar theories, assumes that any item that is overproduced by the economy will have to be worthless. In practice we may find that these items are sold at a great discount, for "practically nothing," but they certainly would not be given away free. We immediately run across obvious over-simplifications in the theory. The act of giving away goods freely is a costly one in a complex civilization. A foundation whose sole purpose is to give away money may allocate up to 15 per cent of its budget for costs of giving the money away. The theory takes no account of these costs.

Again, let us consider the problem of the distribution of goods. In order to demonstrate the existence of an equilibrium, one must assume that the goods produced are divisible in arbitrary amounts. If the product should happen to be a house, it is most peculiar that equilibrium should take place when a fractional number of houses is produced each year. Naturally, one hopes that rounding these numbers off to integers will not materially affect the theory and will present a picture that is close to the actual facts. But even then the question of treating goods is a difficult one. For example, can one treat labor as goods? Or does this violate the facts? Can the theory take the amount of labor and its expense into account the same way as it counts the number of shoes or the amount of steel used in production, or must it somehow be treated as an essentially different item?

This last question gets us into technical, economic questions, which are beyond the scope of this book. However, it was brought up as an example where emotional overtones interfere with the formation of a scientific theory. An economist who tries to treat human labor in the same way that shoes and steel are treated may be ridiculed by his colleagues, or attacked for degrading the subject matter. Of course, it is true that the law of free fall applies equally well to a human being and to a stone, but in the vast majority of cases it is applied to inanimate objects, and the physicists escape this particular difficulty. Here we have an example of the greatest intellectual courage being needed to produce theories without regard to our own prejudices, or to the prejudices of our fellow man.

The greatest danger in a theory, such as the one we are considering, is in the way that it may be used. Suppose that a businessman discovers that the American economy is expanding at a rate of 7 per cent, while the interest rate is only 6 per cent. He may then try to use this theory as a reason why the interest rate should be raised. Of course, the theory says nothing about what *should* be done. It states that, under such-and-such assumptions concerning the behavior of an economy, a certain relation will hold between the expansion rate and the interest rate in equilibrium. If, as a matter of fact, the interest rate is lagging behind the expansion rate, and the economy is in equilibrium, then the assumptions of the theory were incorrect.

Equally dangerous is the question as to whether the economy "should" be in equilibrium. Naturally, this question could be given a precise scientific meaning in terms of human ends commonly derived. However, the chances are that the word "equilibrium" will raise emotional overtones. It suggests stability, and hence a desirable condition. The moment that word enters the scientific literature, the danger is great that social scientists will commit us to working for an equilibrium, without seriously considering the ends to which these means will lead us.

The silliest objection that I have heard to this theory is the following: A social scientist worked through an example of a hypothetical economy and found that the theory did not predict an expansion, but predicted that this economy in equilibrium must be contracting. It was then maintained that a theory that is proposed as a theory of expansion is useless if it sometimes predicts the exact opposite. Here, of course, we are in one of hundreds of verbal pitfalls in using everyday words in the social sciences. The word "expansion" was used by von Neumann to cover expansion, standing still, and contracting. This sort of convenience has been useful in the physical sciences, but has not yet been generally accepted in other fields.

The obvious way out is to replace controversial words by mathematical symbols and conduct the discussion entirely in terms of these symbols. However, this solution has its own inherent dangers. The literature of mathematical applications to the social sciences is full of worthless examples, where intricate mathematical computations have been carried out for quantities that have no application to the world as we know it. The proponents maintain that this is a steppingstone toward the building of a theory of the actual world. The opponents maintain, equally vigorously, that the theory, even if improved, would shed no light on actual applications. This danger is not peculiar to the social sciences but to any science that is just beginning. The question of when a theory is the beginning of great progress and when it is a complete dead end is unanswerable. In the last analysis one must trust to the intuition of the experts in the field and hope that at least some of them are following that rare path which leads to success.

THE FUTURE OF THE SOCIAL SCIENCES

Perhaps a philosopher of Science may allow himself the luxury of crystal-ball-gazing concerning the future of the social sciences. We have seen that the social sciences are just beginning. We have discussed many methodological difficulties in the way of progress in the social sciences. Yet we have felt that there is no essential difference in the scientific method as applicable in the physical sciences, in Biology, Psychology, or in the social sciences. We have noted that the social sciences have the peculiar difficulty of constantly running across ethical questions, which can serve as a great hindrance to progress, but also as a great motivating force. What can we guess as to the future of this humanly all-important field?

The history of Science indicates that most of the avenues explored in the early stages of a branch will lead nowhere. Since the social sciences are more complex than any branch previously attacked by the scientist, we have every reason to believe that most, if not all, present theories will lead nowhere. But this by no means indicates that these undertakings are worthless. Without these false beginnings we would never find a fruitful approach.

We already see many examples where the difficulties in the social sciences correspond to shortcomings in our knowledge of

Mathematics. Therefore, the history of social science is bound to be paralleled by great progress in the development of Mathematics. Indeed, there are distinguished mathematicians who believe that the inspiration that Mathematics received from the physical sciences has nearly come to its end, and that the great new developments in Mathematics will be inspired by problems in the social sciences.

The history of Science certainly indicates that any branch which develops becomes more mathematical. Therefore, the student, who selected the social sciences as a career in the past in order to avoid taking a Mathematics course, may find himself studying more and more Mathematics with the progress of his field. Even now we find specialized journals in various social sciences dealing with the applications of Mathematics, and every indication is that these will grow and eventually take over the entire field. Today an article using mathematical formulas would be looked at askance by many social scientists, while physics has already reached the stage where an article without formulas is held in suspicion. Three hundred years ago physicists still conducted most of their discussions in ordinary language. Today it takes the combined genius of Einstein and Infeld to write a book on modern Physics without formulas. Perhaps the social sciences have just entered this stage.

It is to be expected that for a long time the fruitful results in the social sciences will come in areas of little direct interest to Mankind. But sooner or later results will be obtained from which humans can take guidance for their everyday activities. This will be the stage where the social sciences will have to face their major crisis. It is permissible today for a social scientist to give us advice on what to do, because we don't really feel that he knows what he is talking about. But when this branch reaches a stage where it can make predictions about our actions, and about the results of our actions, with good regularity, human beings will have to face the difficult question of how far the experts are to be trusted in the most basic decisions.

But this stage, too, must pass. Eventually the social sciences will acquire a respectability matching that of the physical sciences. Then human beings who are not experts in a certain field will have to realize that their free decisions lie only in choosing the ends to be achieved. The means are best determined by experts. At this stage an average human being would no more decide whether a certain form of taxation is desirable or not than he would dare pit his own opinion as to the best structure for an airplane against that of the expert engineer.

Without a doubt many of us shudder at the thought of giving up these decisions. However, a by-product of this development is bound to be that we will be forced to think about the ends to be achieved and, perhaps for the first time, be forced to do this in a precise and meaningful form. Ideally the progress of all of Science should reach a stage where the nonscientist would be relieved of all considerations of technical questions, and the sole decisions left up to him would be questions of right and wrong.

SUGGESTED READING

Complete references will be found in the Bibliography at the end of the book.

Neurath.
*Malinowski, pp. 210-227.
Lynd, Chapter V.
Mill [2].
Weber.
Passmore.

16

Quo Vadis?

"Begin at the beginning, and go till you come to the end: then stop."

THAT IS EXCELLENT ADVICE that the King gave to the White Rabbit, but I will not take it. This is the end of the book, but there are still questions to be answered. Where does all this scientific progress lead us? What is the future of Science? What can Science tell us about the future of Mankind? Unfortunately, to answer these questions we need much more information than we now possess. But when questions touch us so deeply, how can we resist guessing?

Just how accurate are our guesses going to be? We know that our information is always only approximate, so a prediction is reliable only if minor errors in our original information do not change the prediction much. For example, if we predict that an atomic-bomb-rocket will hit a certain town, we may consider this prediction reliable. Even if changes in the wind and minor defects in the rocket mechanism carry the rocket half a mile out of its way, the bomb will still hit its target. But if we predict that a card-house will not collapse, then we cannot be certain of this prediction, because a minor error in our data could make the difference between stability and collapse.

We also know that short-range predictions are much more reliable than long-range ones. The reason for this is plain: A minor error will not affect the outcome noticeably in the first few seconds, but it may make all the difference in the long run. When firing a bullet, we can predict where it will be a hundredth of a second later, very accurately. But a minute error (say in the elevation of the rifle), which was not noticeable in so short a time, could make the difference between hitting or missing the target.

Unfortunately, or perhaps fortunately for our peace of mind, the predictions as to our long-range prospects are very unreliable. They are long range, some of the information available is very rough, and many of the theories needed are in their infancy. As an example, let us ask whether the discovery of atomic energy will have a favorable or unfavorable effect on human history in the foreseeable future. Let us try to estimate two opposing factors. Suppose we predict that it will take fifty years to strengthen the UN to the point where it is capable of controlling all weapons (including atomic bombs) effectively. Let us further estimate that there is no danger of an atomic war for the next hundred years. Then we can make a most optimistic prediction, that the atomic age will seem a utopia compared to the present. But if the prediction is that human inertia, the fear of the new in politics, and misguided nationalism prevent the development of the UN for another hundred years, while an atomic war is inevitable within fifty years unless disarmament is enforced before that time, then we can predict a catastrophe.

In spite of the hundreds of articles written in our magazines by "experts," we must confess that we cannot evaluate the factors sufficiently accurately to say which the case is. The social sciences are far from being able to make accurate predictions, the facts are not all known, and very likely the mathematical problems are staggering. Leading social scientists notwithstanding, the best predictions are no more than unfounded guesses; and when we come to such all-important questions, your guess is as good as the expert's.

However, let us assume that mankind will avert the catastrophe, that he will not be wiped out and not even reduced to prehistoric savagery. Let us assume that Science will be given millions of years in which to develop. In that case we are in a position to make very definite predictions. And these predictions will be reliable, because they are based on the predicted trend, not on details. We can certainly foresee a future where men have to work no more than they want to, where every human being has all the necessities of life, and all the luxuries that he cares to have. We can dimly foresee a future in which the vast new sources of energy can power great machines, which can produce a human wealth never dreamed of; a world in which greed may be wiped out, because every wish of every man can be fulfilled.

But this is only a possibility. Our knowledge of Psychology is so limited that no one would dare predict how men will react to these potentialities. Will they work for the goals mentioned, or will they continue to hold it more important that they be "better" than their neighbors? Science will provide all the means necessary for a heaven on earth, but how mankind will use them we cannot tell.

I will not attempt to describe a scientific Utopia. That is the job of our most able science-fiction writers. Let me point out, however, that the invention of the atomic bomb outdid the wildest imagination of the science-fiction writers. We can expect that what these writers are now predicting for the next century will occur in the next decade; what they are predicting for the next decade may already be in production; and what will actually be created within the next century will make *our* whole world look childishly unscientific. Not only is Science progressing, but this progress is getting to be more and more rapid. Newton would have had little trouble grasping the Science of the century after his death, but today every scientist must keep up with the latest periodicals in order not to be hopelessly out of date. There is no question as to whether, say, space-travel will be possible within this century; the question is only what use will be made of this and other developments? Will the new discoveries be used to kill millions and degrade the human stock, or will they be used for the betterment of all mankind?

The revolution that has to come is one in the social sciences, coupled with a careful re-examination of our scale of values. I will not commit the fatal error of telling you that you must re-examine your values, at the same time telling you what the result of this re-examination should be. This is not an essay on morals. But a philosopher of science cannot help observe that the progress of physical science has gotten way ahead of our social progress, both scientific and ethical. And there is every indication that this will continue, unless a concentrated effort is made to promote the study of Man and his goals. The social sciences must be developed to the point where they can present mankind with clear-cut alternatives, which can be debated, and then the course of history can be charted. Today we do not know the alternatives to choose from, and we do not even take the trouble to debate our various courses seriously. But can we be blamed for our neglect? Even if we decided which goal to work for, we would have no idea which path leads to our goal, and which to destruction. The answer to this question belongs to Science, and we can but work as hard as humanly possible to develop the backward relatives of father Physics.

No development of Science alone will solve our great problems, however. Even if we give Man all the means of achieving whatever goals he chooses, the final choice still rests with him. Science can but hope to present him with the alternatives clearly outlined, free from emotional slogans, free from superstitious misrepresentations. But beyond this, Science cannot tell Man what is right and what is wrong. All the progress Science can make throughout human history will be wasted if Man fails to answer the eternal question correctly.

There is one odd fact about long-range predictions—namely, that often the very-long-range predictions are good, even if the long-range ones are not. In the example of the bullet fired, we were unable to predict whether the bullet hit the target, but we can say that it eventually hit the ground (whether it went through the target or not). I would not dare to predict what I will wear a year from today, but I am quite certain about my future, say, a hundred years from now. If you ask me whether a certain atom of uranium will split today, I will be in doubt, but give me a billion billion years, and I am certain that it will have split. Thus Science seems to be able to tell us a great deal about the end of the earth and the universe, even if it can tell us little about the history of the next century.

We are told that there are a number of different ways in which the earth can die, and all of these are likely to happen within the next thirty billion years. The moon can strike the earth, the sun can cool off and freeze the earth, or it can explode, tearing us to bits. All of these predictions ignore, however, the possibility of human interference. I grant that at the moment we see no way of avoiding any of these catastrophies, but who can foresee the progress of Science in billions of years? At any rate there is always the possibility of the human race escaping from the solar system.

There is, however, a gloomier side to the presently accepted theories—the prediction that the whole universe will end. Just as we believe that the universe started when an unimaginably dense and hot gas exploded (this explosion is still going on in the form of the expansion of the universe), we believe that the universe will someday begin contracting again, until it reaches the original stage, wiping out all traces of everything; or, according to an alternate theory, it will keep expanding, leading to a frozen universe. I hope that you have learned one lesson from this book: No scientific theory is final. Tomorrow we may find a new, better theory, which shows us that the present theories are only approximately right. They are excellent approximations for the short span of a million years, but beyond that deviations show up and the net effect may be that the universe will *not* come to an end. I firmly believe that there are some questions, including those I am now discussing, whose answers we shall never attain. Perhaps we were not meant to know the answers.

At the beginning of this book I pointed out the danger of the philosopher trying to usurp the job of the scientist. Now I am in the equally great danger of competing with the prophets. Let me instead reaffirm the philosopher's faith that an increase in knowledge is intrinsically good, and hope that philosophers will continue in their role as critics and as question-raisers to stimulate the unending progress of Science.

BIBLIOGRAPHY

This is in no sense a complete bibliography, but simply a summary of books appearing on the suggested reading lists. The dates of publication given are not necessarily those of the first edition, but rather of the edition relatively most easily available.

The following collections of articles are referred to throughout the bibliography by the indicated abbreviations:

[FS] "Readings in Philosophical Analysis," edited by H. Feigl and W. Sellars, Appleton-Century-Crofts, Inc., New York, 1949.

[FB] "Readings in the Philosophy of Science," edited by H. Feigl and M. Brodbeck, Appleton-Century-Crofts, Inc., New York, 1953.

[SH] "Readings in Ethical Theory," edited by W. Sellars and J. Hospers, Appleton-Century-Crofts, Inc., New York, 1952.

[W] "Readings in the Philosophy of Science," edited by P. P. Wiener, Charles Scribner's Sons, New York, 1953.

[E] "International Encyclopaedia of Unified Science," The University of Chicago Press, Chicago

[LLP] "Albert Einstein Philosopher-Scientist," edited by P. A. Schilpp, Library of Living Philosophers, Tudor Publishing Co., New York, 1951.

Abbott, E. A. "Flatland," Blackwell, Oxford, 1944.

Arley, N. and Buck, K. R. "Introduction to the Theory of Probability and Statistics," John Wiley & Sons, Inc., New York, 1950.

Ayer, A. J. "Critique of Ethics," in [SH].

Beardsley, M. C. "Practical Logic," Prentice-Hall, Inc., New York, 1950.

Benjamin, A. C. "An Introduction to the Philosophy of Science," The Macmillan Co., New York, 1937.

Black, M. "Critical Thinking," Prentice-Hall, Inc., New York, 1952.

Bridgman, P. W. "The Logic of Modern Physics," The Macmillan Co., New York, 1927.

Broad, C. D.

[1] "The Mind and Its Place in Nature," Harcourt, Brace & Co., New York, 1925.

[2] "Review of Julian Huxley's *Evolutionary Ethics*" in [FS].
[3] "Some of the Main Problems of Ethics," in [FS].

Campbell, N. R. "What is Science?" Dover Publications, Inc., New York, 1952.

Carnap, R.
[1] "Foundations of Logic and Mathematics," in [E].
[2] "Logical Foundations of Probability," University of Chicago Press, Chicago, 1950.
[3] "Testability and Meaning," *Philosophy of Science,* October 1936 and January 1937.
[4] "Formal and Factual Science," in [FB].
[5] "The Interpretation of Physics," in [FB].

Cassirer, E. "The Problem of Knowledge," Yale University Press, New Haven, 1950.

Castell, A. "An Introduction to Modern Philosophy," The Macmillan Co., New York, 1943.

Cohen, M. R. and Nagel, E. "An Introduction to Logic and Scientific Method," Harcourt, Brace & Co., New York, 1934.

Conant, J. B. "Science and Common Sense," Yale University Press, New Haven, 1951.

Courant, R. and Robbins, H. "What is Mathematics?" Oxford University Press, New York, 1941.

Darwin, C. R. "The Origin of Species," in [W].

Doyle, Sir A. C. "A Treasury of Sherlock Holmes," Hanover House, Garden City, New York, 1955.

Duhem, P.
[1] "Physical Theory and Experiment," in [FB].
[2] "Representation Versus Explanation," in [W].

Eddington, A. S. "The Nature of the Physical World," The Macmillan Co., New York, 1929.

Einstein, A. "The Laws of Science and the Laws of Ethics," in [FB].

Einstein, A. and Infeld, L. "The Evolution of Physics," Simon and Schuster, Inc., New York, 1942.

Feigl, H.
[1] "Operationism and Scientific Method," in [FS].
[2] "Some Remarks on the Meaning of Scientific Explanation," in [FS].
[3] "The Mind-Body Problem in the Development of Logical Empiricism," in [FB].

Feller, W. "An Introduction to Probability Theory and Its Applications," John Wiley & Sons, Inc., New York, 1958.

Frank, P.
[1] "Einstein, Mach, and Logical Positivism," in [LLP].
[2] "Modern Science and Its Philosophy," Harvard University Press, Cambridge, Mass., 1950.

Frankena, W. K. "The Naturalistic Fallacy," in [SH].

Gamow, G.

[1] "One, Two, Three . . . Infinity," The Viking Press, New York, 1947.

[2] "The Evolutionary Universe," *Scientific American,* September 1956.

Hayakawa, S. I. "Language in Thought and Action," Harcourt, Brace & Co., New York, 1949.

Hempel, C. G.

[1] "Fundamentals of Concept Formation in Empirical Science," in [E].

[2] "Geometry and Empirical Science," in [FS] and in [W].

[3] "On the Nature of Mathematical Truth," in [FS] and in [FB].

Hempel, C. G. and Oppenheim, P. "The Logic of Explanation," in [FB].

Hull, C. L. "Value, Valuation, and Natural Science Methodology," *Philosophy of Science,* July 1944.

Hume, D. "An Enquiry Concerning Human Understanding," Open Court, La Salle, Ill., 1946.

Huxley, J. S. "Evolutionary Ethics," Oxford University Press, New York, 1943.

Kemeny, J. G.

[1] "Man Viewed as a Machine," *Scientific American,* April 1955.

[2] "The Use of Simplicity in Induction," *The Philosophical Review,* July 1953.

Kemeny, J. G. and Oppenheim, P. "On Reduction," *Philosophical Studies,* January-February 1956.

Kemeny, J. G., Snell, J. L. and Thompson, G. L. "Introduction to Finite Mathematics," Prentice-Hall, Inc., Englewood Cliffs, N.J., 1957.

Kemeny, J. G., Mirkil, H., Snell, J. L., and Thompson, G. L., "Finite Mathematical Structures," Prentice-Hall, Inc., Englewood Cliffs, New Jersey, 1959.

Laslett, P., editor. "The Physical Basis of Mind," Blackwell, Oxford, 1950.

Lenzen, V. F. "Procedures of Empirical Science," in [E].

Lewis, C. I. "An Analysis of Knowledge and Valuation," Open Court, La Salle, Ill., 1947.

Lynd, R. S. "Knowledge for What?" Princeton University Press, Princeton, N.J., 1948.

Malinowski, B. "Magic, Science, and Religion," Beacon Press, Inc., Boston, 1948.

Margenau, H. "The Nature of Physical Reality," McGraw-Hill Book Co., Inc., New York, 1929.

Mill, J. S.
[1] "System of Logic," Longmans, Green & Co., Inc., New York, 1929.
[2] "On the Logic of the Social Sciences," in [W].
Mises, R. von. "Positivism," Harvard University Press, Cambridge, Mass., 1951.
Moore, G. E. "Ethics," Oxford University Press, New York, 1949.
Nagel, E.
[1] "Mechanistic Explanation and Organismic Biology," *Philosophy and Phenomenalistic Research,* March, 1951.
[2] "Principles of the Theory of Probability," in [E].
[3] "Teleological Explanation and Teleological Systems," in [FB].
Neurath, O. "Foundations of the Social Sciences," in [E].
Northrop, F. S. C. "Einstein's Conception of Science," in [LLP].
Pap, A. "Elements of Analytical Philosophy," The Macmillan Co., New York, 1949.
Passmore, J. A. "Can the Social Sciences Be Value-Free?" in [FB].
Planck, M. "The Universe in the Light of Modern Physics," Allen and Unwin, London, 1937.
Poincaré, H. "Non-Euclidean Geometries and the Non-Euclidean World," in [FB].
Quine, W. V. "Two Dogmas of Empiricism," *Philosophical Review,* January, 1951.
Reichenbach, H.
[1] "The Logical Foundations of the Concept of Probability," in [FS] and in [FB].
[2] "On the Justification of Induction," in [FS].
Richards, I. A.
[1] "Basic English and Its Uses," W. W. Horton, New York, 1943.
[2] "The Philosophy of Rhetoric," Oxford University Press, New York, 1936.
Robertson, H. P. "Geometry as a Branch of Physics," in [LLP].
Russell, B. "Human Knowledge, Its Scope and Limits," Simon and Schuster, Inc., New York, 1948.
Ryle, G. "The Concept of Mind," Hutchinsons University Library, New York, 1949.
Schlick, M. "Description and Explanation," in [W].
Shaw, G. B. "Back to Methuselah, a Metabiological Pentateuch," Oxford University Press, New York, 1947.
Simpson, G. G. "The Meaning of Evolution," Mentor Books, New York, 1951.
Stevenson, C. L. "Ethics and Language," Yale University Press, New Haven, 1944.

Sullivan, J. W. N. "Limitations of Science," Mentor Books, New York, 1949.

University of California Associates, "The Freedom of the Will," in [FS].

Weber, M. "Objectivity in Social Sciences," in [W].

Werkmeister, W. H. "A Philosophy of Science," Harper & Brothers, New York, 1940.

Whitehead, A. N. "The Abstract Nature of Mathematics," in [W].

Wiener, N. "What is Cybernetics?" in [W].

Wilder, R. L. "Introduction to the Foundations of Mathematics," John Wiley & Sons, Inc., New York, 1952.

Index